SpringerBriefs in Computer Science

For further volumes:
http://www.springer.com/series/10028

Luis Puig · J. J. Guerrero

Omnidirectional Vision Systems

Calibration, Feature Extraction and 3D Information

 Springer

Luis Puig
J. J. Guerrero
University of Zaragoza
Zaragoza
Spain

ISSN 2191-5768   ISSN 2191-5776   (electronic)
ISBN 978-1-4471-4946-0 ISBN 978-1-4471-4947-7   (eBook)
DOI 10.1007/978-1-4471-4947-7
Springer London Heidelberg New York Dordrecht

Library of Congress Control Number: 2013930015

Printed on acid-free paper

Springer is part of Springer Science+Business Media (www.springer.com)

Preface

In this book, we focus on central catadioptric systems. We go from the very early step of calibration to the high-level task of 3D information retrieval. In between we also consider the intermediate step of developing properly adapted features for central catadioptric systems and the analysis of two-view relations between central catadioptric systems and conventional cameras. In the following paragraphs, we describe in more detail the chapters contained in this book.

The model selected to deal with the central catadioptric systems is the sphere camera model, since it gives more information about the elements of the system than other models. In particular it gives important information about the mirror shape used in the catadioptric system. We introduce this model in detail in Chap. 1 along with an analysis of the relation between this model and actual central catadioptric systems. We also introduce the so-called *lifted coordinates* that are used to understand the two-view relations between central catadioptric systems. In Chap. 2 we use the complete general theoretic projection matrix that considers all central catadioptric systems, presented by (Sturm and Barreto 2008), to construct a new approach to calibrate any single-viewpoint catadioptric camera. This approach requires twenty 3D-2D correspondences to compute the projection matrix and the 3D points must lie on at least three different planes. The projection of 3D points on a catadioptric image is performed *linearly* using a 6×10 projection matrix, which uses *lifted coordinates* for image and 3D points. From this matrix, an initial estimation of the intrinsic and extrinsic parameters of the catadioptric system is obtained. We use these parameters to initialize a nonlinear optimization process. This approach is also able to calibrate slightly non-central cameras, in particular, fish-eye cameras. Since reprojection error is not sufficient to determine the accuracy of the approach, we decide to perform a 3D reconstruction from two omnidirectional images.

During the development of our calibration algorithm we realize that there was a lack of deep analysis and comparison of the existing calibration methods for central catadioptric systems. Moreover, for the robotics community where most tasks require to recover information from the environment, calibration of cameras is a basic step. At the same time this step should be easy to perform and reliable. In this order we perform a classification of the existing approaches. On the other hand, we select those approaches which are already available as OpenSource and which do not require a complex pattern or scene to perform a comparison using synthetic and real images, so the user could select the more convenient for its particular case. In Chap. 3 we present these methods and an analysis of their advantages and drawbacks.

In Chap. 4 we present a deep analysis of the two-view relations of uncalibrated central catadioptric systems. We particularly pay attention to the mixture of central catadioptric systems and perspective cameras, which we call *hybrid*. The two-view geometric relations we consider are the hybrid fundamental matrix and the hybrid planar homography. These matrices contain useful geometric information. We study three different types of matrices, varying in complexity depending on their capacity to deal with a single or multiple types of central catadioptric systems. The first and simplest one is designed to deal with paracatadioptric systems, the second one and more complex, considers the combination of a perspective camera and any central catadioptric system. The last one is the complete and generic model which is able to deal with any combination of central catadioptric systems. We show that the generic and most complex model sometimes is not the best option when we deal with real images. Simpler models are not as accurate as the complete model in the ideal case, but they provide a better and more accurate behavior in presence of noise, being simpler and requiring less correspondences to be computed. Finally, using the best models we present the successful matching between perspective images and hypercatadioptric images introducing geometrical constraints into a robust estimation technique.

Another basic step in vision and robotics applications is feature detection/extraction. Through the years several techniques have been developed for conventional (perspective) cameras. The SIFT proposed by (Lowe 2004) has become the most used feature extraction approach. This scale-invariant approach is based on the approximation to the Laplacian of Gaussians (LoG) through the difference of Gaussians (DoG). The Gaussian filtering in Euclidean computer vision, which is required to construct the scale space of the images can be computed in two ways: either using convolution with a Gaussian kernel or by implementing the linear heat flow. In Chap. 5 we develop a new approach to compute the scale space of any omnidirectional image acquired with a central catadioptric system. We combine the sphere camera model and the partial differential equations framework on manifolds, to compute the Laplace–Beltrami (LB) operator which is a second-order differential operator required to perform the Gaussian smoothing on catadioptric images. Finally in Chap. 6 we present an approach to compute the orientation of a hand-held omnidirectional catadioptric camera. We use the vanishing points which contain geometric information related to the orientation of the catadioptric system with respect to the dominant directions in man-made environments. The vanishing points are computed from the intersection of parallel lines. The 3D lines are projected in catadioptric images as conics. We extract analytically the projected lines corresponding to straight lines in the scene by using the internal calibration and two image points that lie on the corresponding line projection.

Contents

Chapter 1
Modeling Omnidirectional Vision Systems

Abstract In this chapter, different types of omnidirectional systems are briefly introduced. Then, we focus on the central catadioptric systems and the model used to deal with this type of systems, the so-called sphere camera model. The projection of points and lines under this model are also explained as well as the relation between this model and the actual catadioptric systems. Later, we introduce the lifted coordinates, which is a tool used to deal with the non linearities present on the sphere camera model. We show two different forms to compute them. The former makes use of the G operator and the latter one uses symmetric matrix equations. Finally, a useful representation of catadioptric systems as Riemannian manifolds is presented.

1.1 Introduction

In recent years, the use of omnidirectional cameras has widely increased between the computer vision and robotics communities. The major advantage of this type of cameras is their wide field of view (FOV) which allows them to include the whole scene in a single view. They have been used in such different areas as surveillance, tracking, visual navigation, localization and SLAM, structure from motion, active vision, visual odometry, photogrammetry, camera networks, reconstruction of cultural heritage, and so on. There exist several types of omnidirectional cameras which can be classified as central and noncentral. Among the noncentral cameras we can find the rotating camera, which consists of a conventional camera with a mechanic system that allows it to move along a circular trajectory and to acquire images from the surroundings. Polycameras which are camera clusters of conventional cameras pointing to different directions in a particular configuration. Another type of omnidirectional systems is dioptric systems which use wide-angle lenses such as fish-eye lenses combined with conventional cameras. On the other hand, central omnidirectional cameras are those which satisfy the single-viewpoint property. This is an important property since it allows to easily calculate the directions of light rays coming into the camera. Baker and Nayar (1999) extensively studied the catadioptric

L. Puig and J. J. Guerrero, *Omnidirectional Vision Systems*, SpringerBriefs in Computer Science, DOI: 10.1007/978-1-4471-4947-7_1, © Luis Puig 2013

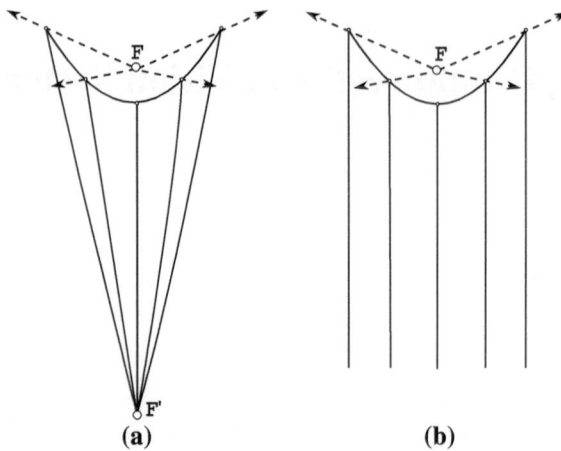

Fig. 1.1 Mirrors used in configurations with conventional cameras. **a** Hyperbolic. **b** Parabolic

systems, combinations of camera lenses and mirrors. They proved that the elliptic, parabolic, and hyperbolic mirrors, combined with conventional cameras, are the only ones that ensure the single-viewpoint property, provided that the mirror is positioned appropriately relative to the camera. The two most popular of such systems are the hypercatadioptric system and the paracatadioptric system. The former is composed by a hyperbolic mirror (Fig. 1.1a) and a perspective camera. The latter is composed by a parabolic mirror (Fig. 1.1b) and an orthographic camera. Including fish-eye lenses, these are the three omnidirectional systems most used by the computer vision and robotics communities. Two omnidirectional cameras are shown in Fig. 1.2.

There exist several geometric and analytic models to deal with omnidirectional systems, see Sturm et al. (2011). In the case of central catadioptric systems, Svoboda and Pajdla (2002) propose different models for different mirrors and give formulae for the associated epipolar geometry. Strelow et al. (2001) deal directly with the reflection properties of the rays on the mirror. A unified model was proposed by Geyer and Daniilidis (2000), where they present the sphere camera model which allows to deal with any central catadioptric system. Later this model was extended by Barreto and Araujo (2001) and Ying and Hu (2004a). Recently, another improvement to this model, which completes the model of central catadioptric systems, was presented by Sturm and Barreto (2008). They explain the generic projection matrix, as well as the general 15×15 fundamental matrix, and plane homographies. This model is one of the most used models in current days, since it provides important information about the mirror shape. This information can be used in different tasks such as calibration, line detection, computing geometrical invariants, computing the scale space for a particular mirror, and so on. With respect to the slightly noncentral systems, in particular for fish-eye lenses we can find the following approaches. Swaminathan and Nayar (1999) model this type of projections as a combination of three types of distortion. These distortions are the shift of the optical center, radial distortion,

Fig. 1.2 Examples of omnidirectional cameras. **a** Catadioptric system. **b** Fisheye lens

(a)   (b)

and decentering distortion. Micusik and Pajdla (2003) compute the projection of 3D points to the camera plane using trigonometric functions which are linearized through Taylor series. This is done for a particular type of camera. Kannala and Brandt (2004) propose a generic model to deal with all cameras equipped with fish-eye lenses. They consider the projections as a series of odd powers of the angle between the optical axis and the incoming ray, then they complete the model by adding radial and tangential distortion. Courbon et al. (2007) propose a generic model to calibrate any fish-eye system based on the sphere camera model. Another category that should be mentioned is the generic methods that can model any arbitrary imaging system. Grossberg and Nayar (2001) propose a method based on virtual sensing elements called raxels which describe a mapping from incoming scene rays to photo-sensitive elements on the image detector. This work has inspired many works and a list of some of them can be found in Ramalingam et al. (2005).

In this book we model the central catadioptric systems, which are a composition of cameras and mirrors, by using the sphere camera model, proposed originally by Geyer and Daniilidis (2000) and later improved by Barreto and Araujo (2001). Recently, a general model was presented in Sturm and Barreto (2008), which proves the existence of a bilinear matching constraint for all central catadioptric cameras. In this chapter, we present the projection of points and lines under the more recent and complete model. On the other hand, we explain the relation between the sphere camera model and the actual central catadioptric systems with its constructive mirror, optics, and camera parameters. Another useful tool to deal with the central catadioptric systems, and more specific with the nonlinearities introduced by the sphere camera model, are the *lifted coordinates*. Barreto and Daniilidis (2006) use Veronese maps to create these *lifted coordinates*. They can represent in a single entity, the conic, the two image points generated by the projection of a 3D point under the sphere camera model. Moreover, because of the dual principle, they can also represent two lines using the same entity, the conic. These *lifted coordinates* are particularly useful when we explain the two-view relations between central catadioptric systems and the general projection under the sphere camera model. Another relevant tool to deal with central catadioptric systems are the partial differential equations, since the mirrors of the catadioptric systems can be seen as parametric surfaces. These equations will allow us to compute the metric of corresponding mirror which encodes its geometrical properties. These three elements are basic tools for the development of the majority of the contributions of the following book, so they are presented in this initial background chapter.

1.2 Sphere Camera Model

The sphere camera model used to explain central catadioptric systems was initially introduced by Geyer and Daniilidis (2000). All central catadioptric cameras can be modeled by a unit sphere and a perspective projection, such that the projection of 3D points can be performed in two steps (Fig. 1.3). First, one projects the point onto the unit sphere, obtaining the intersection of the sphere and the line joining its center and the 3D point. There are two such intersection points, which are represented as $\mathbf{s}_\pm \sim (Q_1, Q_2, Q_3, \pm\sqrt{Q_1^2 + Q_2^2 + Q_3^2})^\mathsf{T}$. These points are then projected in the second step, using a perspective projection P resulting in two image points $\mathbf{q}_\pm \sim \mathsf{P}\mathbf{s}_\pm$, one of which is physically true. This model covers all central catadioptric cameras, encoded by ξ, which is the distance between the center of the perspective

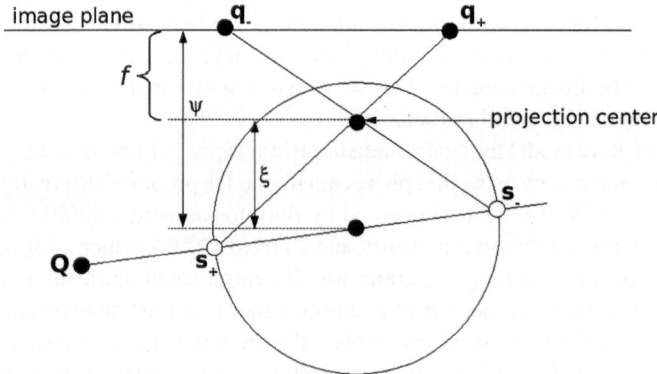

Fig. 1.3 Projection of a 3D point to two image points in the sphere camera model. The z-axis of the camera coordinate system is positive upwards. The camera is looking up

projection and the center of the sphere, and ψ which is the distance between the center of the sphere and the image plane. We have $\xi = 0$ for perspective, $\xi = 1$ for paracatadioptric and $0 < \xi < 1$ for hypercatadioptric cameras.

Let the unit sphere be located at the origin and the optical center of the perspective camera, at the point $\mathbf{C}_p = (0, 0, \xi)^{\mathsf{T}}$. The perspective camera is modeled by the projection matrix $\mathsf{P} \sim \mathsf{A}_p\mathsf{R}_p(\mathsf{I} - \mathbf{C}_p)$, where A_p is its calibration matrix. We assume it is of the form

$$\mathsf{A}_p = \begin{pmatrix} f & 0 & c_x \\ 0 & f & c_y \\ 0 & 0 & 1 \end{pmatrix} \tag{1.1}$$

with f the focal length and (c_x, c_y) the principal point. The rotation R_p denotes a rotation of the perspective camera looking at the mirror (this rotation is usually very small, thus often neglected). Rotation about the z-axis can always be neglected since it is coupled with the rotation of the whole system about the z-axis. Since both intrinsic and extrinsic parameters of the *perspective* camera are intrinsic parameters for the *catadioptric* camera, we replace $\mathsf{A}_p\mathsf{R}_p$ by a generic projective transformation K. Note that the focal length of the perspective camera in the sphere model is different from the focal length of the physical camera looking at the mirror; its value is actually determined by the physical camera's focal length f_c, the mirror parameters (ξ, ψ) and the rotation between the camera and the mirror (R_p). More explicitly, the projections of the points on the sphere $\mathbf{s}_\pm$ to points on the omnidirectional image are obtained as

$$\mathbf{q}_\pm \sim \mathsf{P}\mathbf{s}_\pm, \quad \text{where} \quad \mathsf{P} \sim \mathsf{K} \left(\mathsf{I} \ \middle| \ \begin{matrix} 0 \\ 0 \\ -\xi \end{matrix} \right), \tag{1.2}$$

with I, a 3×3 identity matrix. Giving the final definition of

$$\mathbf{q}_\pm \sim \mathsf{K} \begin{bmatrix} \mathcal{Q}_1 \\ \mathcal{Q}_2 \\ \mathcal{Q}_3 \pm \xi\sqrt{\mathcal{Q}_1^2 + \mathcal{Q}_2^2 + \mathcal{Q}_3^2} \end{bmatrix}. \tag{1.3}$$

To simplify, it is usual to work with the intermediate image points $\mathbf{r}_\pm \sim \mathsf{K}^{-1}\mathbf{q}_\pm$. Explicitly defined as $\mathbf{r}_\pm = (\mathcal{Q}_1, \mathcal{Q}_2, \mathcal{Q}_3 \pm \xi\sqrt{\mathcal{Q}_1^2 + \mathcal{Q}_2^2 + \mathcal{Q}_3^2})^{\mathsf{T}}$, before giving final results for the actual image points $\mathbf{q}_\pm$. The theoretical two image points $\mathbf{q}_\pm$ can be represented in a single geometric object, which is the degenerate dual conic generated by the two points. This conic contains all lines incident to either one or both of these two points $\Omega \sim \mathbf{q}_+\mathbf{q}_-^{\mathsf{T}} + \mathbf{q}_-\mathbf{q}_+^{\mathsf{T}}$.

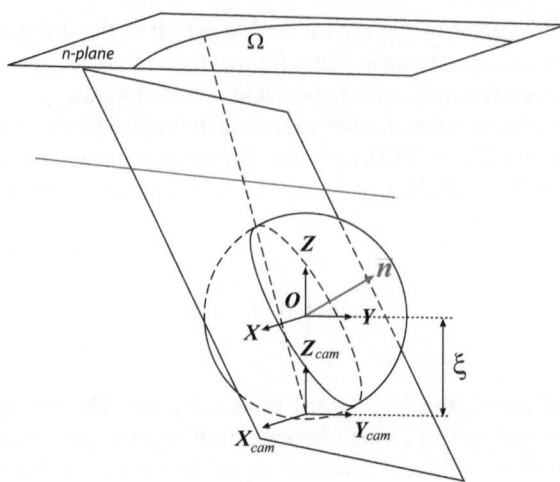

Fig. 1.4 Projection of a line under the sphere camera model

1.2.1 Projections of Lines

The projections of lines under the sphere camera model are explained as follows. Let $\Pi = (n_x, n_y, n_z, 0)^\mathsf{T}$ a plane defined by a 3D line and the effective view point in the sphere camera model $\mathbf{O}$ (see Fig. 1.4). The 2D line $\mathbf{n}$ associated to the 3D line by P can be represented as $\mathbf{n} = (n_x, n_y, n_z)^\mathsf{T}$. Then, the points $\mathbf{Q}$ lying on the 3D line are projected to points $\mathbf{q}$. These points satisfy $\mathbf{n}^\mathsf{T}\mathbf{q} = 0$ and $\mathbf{q} = \mathsf{K}\mathbf{r}$, so $\mathbf{n}^\mathsf{T}\mathsf{K}\mathbf{r} = 0$. This equality can be written as

$$\mathbf{r}^\mathsf{T}\Omega\mathbf{r} = 0 \tag{1.4}$$

where the image conic is

$$\Omega = \begin{pmatrix} n_x^2\left(1 - \xi^2\right) - n_z^2\xi^2 & n_x n_y\left(1 - \xi^2\right) & n_x n_z \\ n_x n_y\left(1 - \xi^2\right) & n_y^2\left(1 - \xi^2\right) - n_z^2\xi^2 & n_y n_z \\ n_x n_z & n_y n_z & n_z^2 \end{pmatrix} \tag{1.5}$$

and the image of the conic in the catadioptric image is

$$\widehat{\Omega} = \mathsf{K}^{-\mathsf{T}}\Omega\mathsf{K}^{-1}. \tag{1.6}$$

Notice that Ω is a degenerate conic when the 3D line is coplanar with the optical axis (Barreto and Araujo 2005).

1.2.2 Relation Between the Real Catadioptric System and the Sphere Camera Model

Here, we analyze the relation between the parameters present in a real catadioptric system and their representation in the sphere camera model. We recover the intrinsic parameters of the real catadioptric system from their counterparts in the sphere camera model. We also analyze the tilting and focal length f_c of the conventional camera looking at the mirror.

The focal length in the sphere model is not the same as the focal length of the real camera, looking at the mirror. This is best seen for the paracatadioptric case, where the real camera is orthographic (infinite focal length), whereas the perspective camera in the sphere camera model has a finite focal length. The analogous is also valid for tilting parameters.

1.2.2.1 Tilting

Tilting in a camera can be defined as a rotation of the image plane with respect to the pinhole. This is also equivalent to tilting the incoming rays since both have the same pivoting point: the pinhole. In the Fig. 1.5 the tilt in a catadioptric camera is represented. Similarly, the tilt in the sphere model (R_p in $K = A_p R_p$) corresponds to tilting the rays coming to the perspective camera of the sphere model (Fig. 1.6). Although the same image is generated by both models, the tilting angles are not identical, even they are not proportional to each other. So, it is also not possible to obtain the real system tilt amount by multiplying the sphere model tilt by a coefficient.

1.2.2.2 Focal Length f

The composition of paracatadioptric and hypercatadioptric systems is different. The first one uses a parabolic mirror and an orthographic camera. In this case the focal length of the real system, f_c, is infinite.

For the hypercatadioptric system, we are able to relate f with the focal length of the perspective camera in the real system, f_c. We start defining explicitly the projection matrix K. Assuming image skew is zero, $R_p = I$ and principal point is $(0, 0)$, K is given in Barreto and Araujo (2005) as

$$K = \begin{pmatrix} (\psi - \xi)f_c & 0 & 0 \\ 0 & (\psi - \xi)f_c & 0 \\ 0 & 0 & 1 \end{pmatrix} \quad (1.7)$$

where ψ corresponds to the distance between the effective viewpoint and the re-projection plane (cf. Fig. 1.3). The relation between the focal lengths is $f = (\psi - \xi)f_c$. From the same reference Barreto and Araujo (2005) we get

Fig. 1.5 Tilt in a central catadioptric system

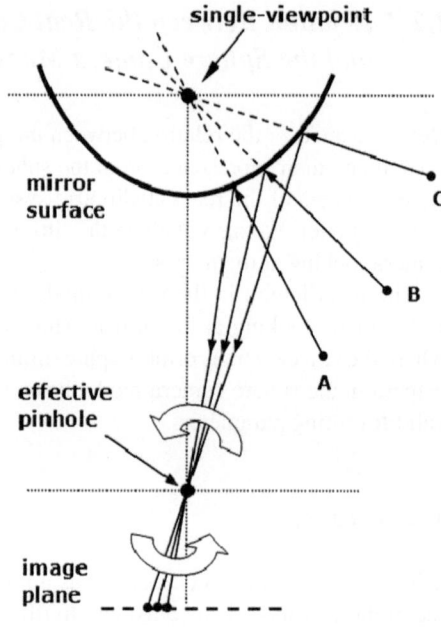

Fig. 1.6 Tilt in the sphere camera model

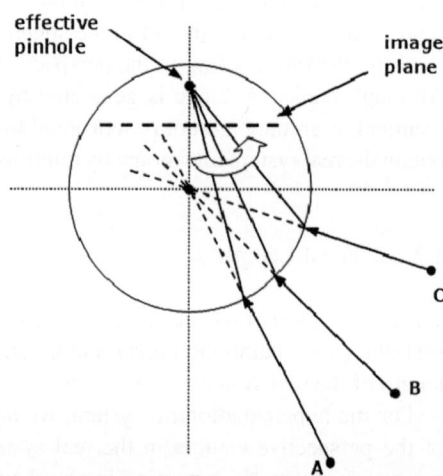

$$\xi = \frac{d}{\sqrt{d^2 + 4p^2}} \qquad \psi = \frac{d + 2p}{\sqrt{d^2 + 4p^2}} \qquad (1.8)$$

where d is the distance between the foci of the hyperboloid and $4p$ equals to the latus rectum. Developing the equations we obtain p in terms of d and ξ, $2p = d\sqrt{1 - \xi^2}/\xi$, which is used to obtain $\psi = \xi + \sqrt{1 - \xi^2}$. With this final relation we can write

$$f = f_c\sqrt{1 - \xi^2} \tag{1.9}$$

from which we extract the focal length of the perspective camera in the real system

$$f_c = \frac{f}{\sqrt{1 - \xi^2}}. \tag{1.10}$$

1.3 Lifted Coordinates and Matrices

The derivation of (multi-)linear relations for catadioptric imagery requires the use of lifted coordinates . They allow to generalize the transformations and multiview tensors from conventional perspective images to catadioptric systems, where the projective invariant entities are quadrics instead of lines.

The Veronese map $V_{n,d}$ of degree d maps points of $\mathbb{P}^n$ into points of an m dimensional projective space $\mathbb{P}^m$, with $m = \binom{n+d}{d} - 1$. Consider the second-order Veronese map $V_{2,2}$, that embeds the projective plane into the 5D projective space, by lifting the coordinates of point $\mathbf{q} = (q_1, q_2, q_3)^\mathsf{T}$ to

$$\hat{\mathbf{q}} = (q_1^2, \, q_1 q_2, \, q_2^2, \, q_1 q_3, \, q_2 q_3, \, q_3^2)^\mathsf{T}. \tag{1.11}$$

This lifting preserves homogeneity and it is suitable to deal with quadratic functions, because it discriminates the entire set of second-order monomials (Barreto and Daniilidis 2006).

As we observe, if $\mathbf{c} = (c_1, c_2, c_3, c_4, c_5, c_6)^\mathsf{T}$ represents a conic, its equation $c_1 q_1^2 + c_2 q_1 q_2 + c_3 q_2^2 + c_4 q_1 q_3 + c_5 q_2 q_3 + c_6 q_3^2 = 0$, can be written as $\hat{\mathbf{q}}^\mathsf{T}\mathbf{c} = 0$.

When the conic $\mathbf{c}$ has the particular shape of a circle, we have $c_2 = 0$ and $c_1 = c_3$. We then use the simplified lifted coordinates of a point $\mathbf{q} = (q_1, q_2, q_3)$ in a 4-vector defined as

$$\hat{\mathbf{q}} = (q_1^2 + q_2^2, \, q_1 q_3, \, q_2 q_3, \, q_3^2)^\mathsf{T}. \tag{1.12}$$

There are two ways to compute this lifting, which not only involves vectors but matrices. The first one is presented by Barreto and Daniilidis (2006) and makes use of the Γ operator. The second one is more general and it is presented by Sturm and Barreto (2008). It makes use of symmetric matrix equations.

1.3.1 Lifted Coordinates and Matrices Using the Γ Operator

The lifting presented in (Barreto and Daniilidis 2006) is limited to transform 3×1 vectors into 6×1 vectors and 3×3 matrices into 6×6 matrices. It makes use of the

Γ operator, which is defined for two points $\mathbf{q}$ and $\bar{\mathbf{q}}$ as follows:

$$\Gamma(\mathbf{q}, \bar{\mathbf{q}}) = \left(q_1\bar{q}_1, \frac{q_1\bar{q}_2 + q_2\bar{q}_1}{2}, q_2\bar{q}_2, \frac{q_1\bar{q}_1 + q_3\bar{q}_1}{2}, \frac{q_2\bar{q}_3 + q_3\bar{q}_2}{2}, q_3\bar{q}_3\right)^{\mathsf{T}}.$$
(1.13)

When this operator is used with the same point we obtain the lifting of its coordinates.

$$\Gamma(\mathbf{q}, \mathbf{q}) = (q_1^2,\ q_1q_2,\ q_2^2,\ q_1q_3,\ q_2q_3,\ q_3^2)^{\mathsf{T}}.$$
(1.14)

With respect to 3×3 matrices, the way the Γ operator is used to perform the lifting is the following. Matrices can be considered as linear transformations. Let us define a linear transformation L, which maps points $\mathbf{x}$ and $\bar{\mathbf{x}}$ to points $\mathsf{L}\mathbf{x}$ and $\mathsf{L}\bar{\mathbf{x}}$, respectively. The operator Υ, that lifts the transformation L from the projective plane $\wp^2$ to the embedding space $\wp^5$, i.e., maps a 3×3 matrix L to a 6×6 matrix $\widehat{\mathsf{L}}$ must satisfy

$$\Gamma(\mathsf{L}\mathbf{x}, \mathsf{L}\bar{\mathbf{x}}) = \Upsilon(\mathsf{L}).\Gamma(\mathbf{x}, \bar{\mathbf{x}})$$
(1.15)

Such operator can be derived by algebraic manipulation. Let $\mathbf{v}_1$, $\mathbf{v}_2$, $\mathbf{v}_3$ be the column vectors of L and $\Gamma_{ij} = \Gamma(\mathbf{v}_i, \mathbf{v}_j)$, the lifted representation of matrix L is

$$\widehat{\mathsf{L}} = \Upsilon(\mathsf{L}) = [\Gamma_{11}\Gamma_{12}\Gamma_{22}\Gamma_{13}\Gamma_{23}\Gamma_{33}]\widetilde{\mathsf{D}}, \quad \text{with} \quad \widetilde{\mathsf{D}} = diag\{1, 2, 1, 2, 2, 1\}.$$
(1.16)

1.3.2 Lifted Coordinates and Matrices Using Symmetric Matrix Equations

This approach used in (Sturm and Barreto 2008) is more general than the previously presented. It can deal with higher dimension vectors, since it is based on symmetric matrix equations. In general terms, it can be described as the lower triangular part of matrix M, which is obtained from the multiplication of the vector $\mathbf{q}$ by its corresponding transpose vector $\mathbf{q}^{\mathsf{T}}$, $\mathsf{M} = \mathbf{q}\mathbf{q}^{\mathsf{T}}$. More specifically, a vector $\hat{\mathbf{q}}$ and matrix $\mathbf{q}\mathbf{q}^{\mathsf{T}}$ are composed by the same elements. The former can be derived from the latter through a suitable re-arrangement of parameters. Define $\mathbf{v}(\mathsf{U})$ as the vector obtained by stacking the columns of a generic matrix U (Horn and Johnson 1991). For the case of $\mathbf{q}\mathbf{q}^{\mathsf{T}}$, $\mathbf{v}(\mathbf{q}\mathbf{q}^{\mathsf{T}})$ has several repeated elements because of the matrix symmetry. By left multiplication with a suitable permutation matrix S that adds the repeated elements, it follows that

$$\hat{\mathbf{q}} = \mathsf{D}^{-1}\mathsf{S}\mathbf{v}(\mathbf{q}\mathbf{q}^{\mathsf{T}}),$$
(1.17)

with D a diagonal matrix, $D_{ii} = \sum_{j=1}^{\|\mathsf{M}\|} S_{ij}$.

If U is symmetric, then it is uniquely represented by $\mathbf{v}_{\text{sym}}(\mathsf{U})$, the row-wise vectorization of its lower left triangular part:

$$\mathbf{v}_{\text{sym}}(\mathsf{U}) = \mathsf{D}^{-1}\mathsf{S}\mathbf{v}(\mathsf{U}) = (U_{11}, U_{21}, U_{22}, U_{31}, \ldots, U_{nn})^{\mathsf{T}} \qquad (1.18)$$

Since S gives us the position of the repeated elements of $\mathbf{v}(\mathsf{U})$, it is easy to recover $\mathbf{v}(\mathsf{U})$ from $\mathbf{v}_{\text{sym}}(\mathsf{U})$

$$\mathbf{v}(\mathsf{U}) = \mathsf{S}^{\mathsf{T}}\mathbf{v}_{\text{sym}}(\mathsf{U}) \qquad (1.19)$$

There are two particular liftings that are useful in the present book. The one we explained in last section, which corresponds to the Veronese map $V_{2,2}$, which maps 3 vectors into 6 vectors. The second one corresponds to the Veronese map $V_{3,2}$, which maps 4 vectors $\mathbf{Q} = (Q_1, Q_2, Q_3, Q_4)^{\mathsf{T}}$ into 10 vectors $\widehat{\mathbf{Q}}$. Using the theory explained above we can define the Veronese map $V_{2,2}$ as

$$\widehat{\mathbf{q}} = \mathbf{v}_{\text{sym}}(\mathbf{q}\mathbf{q}^{\mathsf{T}}) = \begin{pmatrix} q_1^2 & q_1q_2 & q_1q_3 \\ q_1q_2 & q_2^2 & q_2q_3 \\ q_1q_3 & q_2q_3 & q_3^2 \end{pmatrix} = (q_1^2,\, q_1q_2,\, q_2^2,\, q_1q_3,\, q_2q_3,\, q_3^2)^{\mathsf{T}}.$$

$$\qquad (1.20)$$

and the Veronese map $V_{3,2}$ as

$$\widehat{\mathbf{Q}} = \mathbf{v}_{\text{sym}}(\mathbf{Q}\mathbf{Q}^{\mathsf{T}}) = \begin{pmatrix} Q_1^2 & Q_1Q_2 & Q_1Q_3 & Q_1Q_4 \\ Q_1Q_2 & Q_2^2 & Q_2Q_3 & Q_2Q_4 \\ Q_1Q_3 & Q_2Q_3 & Q_3^2 & Q_3Q_4 \\ Q_1Q_4 & Q_2Q_4 & Q_3Q_4 & Q_4^2 \end{pmatrix} = \begin{pmatrix} Q_1^2 \\ Q_1Q_2 \\ Q_2^2 \\ Q_1Q_3 \\ Q_2Q_3 \\ Q_3^2 \\ Q_1Q_4 \\ Q_2Q_4 \\ Q_3Q_4 \\ Q_4^2 \end{pmatrix} \qquad (1.21)$$

Let us now discuss the lifting of linear transformations (matrices), induced by lifting of points. Consider A such that $\mathbf{r} = \mathsf{A}\mathbf{q}$. The relation $\mathbf{r}\mathbf{r}^{\mathsf{T}} = \mathsf{A}(\mathbf{q}\mathbf{q}^{\mathsf{T}})\mathsf{A}^{\mathsf{T}}$ can be written as a vector mapping

$$\mathbf{v}(\mathbf{r}\mathbf{r}^{\mathsf{T}}) = (\mathsf{A} \otimes \mathsf{A})\mathbf{v}(\mathbf{q}\mathbf{q}^{\mathsf{T}}), \qquad (1.22)$$

with $\otimes$ denoting the Kronecker product (Horn and Johnson, 1991). Using the symmetric vectorization, we have $\hat{\mathbf{q}} = \mathbf{v}_{\text{sym}}(\mathbf{q}\mathbf{q}^{\mathsf{T}})$ and $\hat{\mathbf{r}} = \mathbf{v}_{\text{sym}}(\mathbf{r}\mathbf{r}^{\mathsf{T}})$, thus, from (1.19) and (1.22):

$$\hat{\mathbf{r}} = \widehat{\mathsf{A}}\hat{\mathbf{q}} = \mathsf{D}^{-1}\mathsf{S}(\mathsf{A} \otimes \mathsf{A})\mathsf{S}^{\mathsf{T}}\hat{\mathbf{q}} \qquad (1.23)$$

where the 6×6 matrix $\widehat{\mathsf{A}}$, represents the lifted linear transformation.

Finally, let us note that useful properties of the lifting of transformations are given in (Horn and Johnson 1985, 1991):

$$\widehat{AB} = \widehat{A}\widehat{B} \qquad \widehat{A^{-1}} = \widehat{A}^{-1} \qquad \widehat{A^{T}} = D^{-1}\widehat{A}^{T}D. \tag{1.24}$$

1.4 Generic Projection to Build a Metric
for a Riemannian Manifold

The mirrors of central catadioptric systems, which are composed of cameras and mirrors, can be seen as parametric surfaces $\mathcal{M}$ in $\mathbb{R}^3$. Their geometrical properties are encoded in their induced Riemannian metric g_{ij}, for which the partial differential equations are a very versatile tool that is well suited to cope with the computation of such metric.

In the catadioptric systems, any ray incident in one of the foci of the mirror is projected to the other focus. In the case of the paracatadioptric system this focus is located at the infinity. A pixel is created where the reflected light ray intersects the camera plane.

Let (q_1, q_2) and (Q_1, Q_2, Q_3) be coordinates in the image plane, which we consider an open subset $\Xi \subseteq \mathbb{R}^2$, and on the mirror surface, respectively. The image formation process previously described induces a mapping between manifolds from the surface of the mirror to the camera plane

$$\begin{aligned} \Psi : \mathcal{M} &\rightarrow \Xi \\ (Q_1, Q_2, Q_3) &\rightarrow (q_1, q_2) \end{aligned} \tag{1.25}$$

This mapping allows to transport the metric $g_{ij}(Q_1, Q_2, Q_3)$ and to provide the image with a pullback metric $h_{ij}(q_1, q_2)$

$$h_{kl} = \frac{\partial}{\partial q_k} Q_r \frac{\partial}{\partial q_l} Q_s g_{rs}, \qquad k, l \in \{1, 2\}, r, s \in \{0, 1, 2\}, \tag{1.26}$$

the Einstein's convention summation is used. Notice that the image plane (Ξ, h) carries a very specific geometry encoded in its metric h_{ij}. This geometry is inherited from that of the mirror $(\mathcal{M}, g)$, but it uses a more conventional domain $\Xi \subseteq \mathbb{R}^2$. Therefore, with this simpler and geometrically accurate parameterization we are able to perform computations on $\mathcal{M}$ easily.

Omnidirectional images can be treated as scalar fields on parametric manifolds. With this idea Riemannian geometry is used to derive generalizations of concepts of smoothing and scale space for this particular type of images. A general framework is derived based on energy minimization and partial differential equations. The parameterization is used to compute efficiently these operations *directly from the camera plane* and at the same time respecting the complex geometry of the manifold.

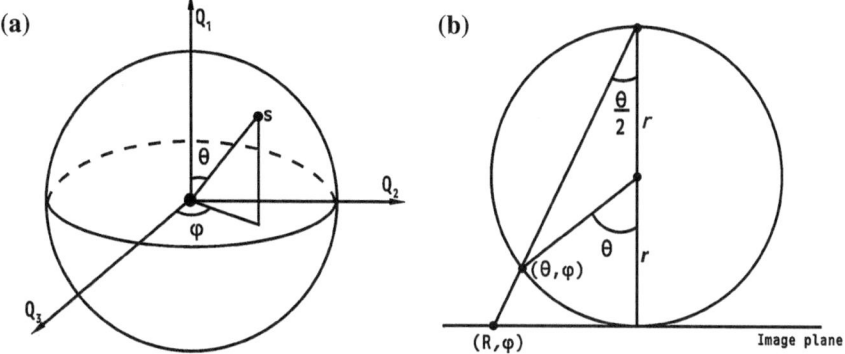

Fig. 1.7 Geometry of the 2-sphere $\mathbb{S}^2$. **a** Spherical polar coordinates. **b** Stereographic projection

1.4.1 Paracatadioptric System: An Example

Geyer and Daniilidis (2001a) showed that the paracatadioptric projection is equivalent to the inverse stereographic projection. In this case we need to introduce another manifold, which corresponds to the 2-dimensional sphere $\mathbb{S}^2$ in $\mathbb{R}^3$. Notice that using this approach not only the paracatadioptric systems can be modeled through a projection on the sphere, but all central projection systems.

We now require to link the geometry of $\mathbb{S}^2$ to the one of the sensor image. This will allow to process the spherical image directly using the sensor output. Consider a sphere of radius r (Fig. 1.7a). A point **s** on $\mathbb{S}^2$ is defined in cartesian and polar coordinates as

$$(s_1, s_2, s_3) = r(\cos\theta, \sin\theta\sin\varphi, \sin\theta\cos\varphi), \quad \text{with} \quad \theta \in [0, \pi), \; \varphi \in [0, 2\pi) \tag{1.27}$$

The Euclidean line element can be represented in cartesian and polar coordinates

$$dl^2 = ds_1^2 + ds_2^2 + ds_3^2 = r^2(d\theta^2 + \sin^2\theta d\varphi^2) \tag{1.28}$$

The stereographic projection (see Fig. 1.7b) sends a point (θ, φ) on the sphere to the point with polar coordinates (R, φ) in the plane, with $\varphi = \varphi$ and $R = 2r\tan\frac{\theta}{2}$. Then the terms in (1.28) are

$$d\theta^2 = \frac{16r^2}{(4R^2 + r^2)^2}dR^2, \quad \sin^2(\theta) = \frac{16r^2}{(R^2 + 4r^2)^2}R^2 \tag{1.29}$$

simplifying

$$dl^2 = \frac{16r^4}{(R^2 + 4r^2)^2}(dR^2 + R^2d\varphi^2) \tag{1.30}$$

Let $(q_1, q_2) \in \mathbb{R}^2$ on the sensor plane define cartesian coordinates, where $R^2 = q_1^2 + q_2^2$, $\varphi = \arctan \frac{q_2}{q_1}$ and $r = 1$. The line element becomes

$$dl^2 = \frac{16}{(4 + q_1^2 + q_2^2)^2} (dq_1^2 + dq_2^2) \tag{1.31}$$

giving the Riemannian metric h_{ij}

$$h_{ij}(q_1, q_2) = \begin{pmatrix} \frac{16}{(4+q_1^2+q_2^2)^2} & 0 \\ 0 & \frac{16}{(4+q_1^2+q_2^2)^2} \end{pmatrix} \tag{1.32}$$

and its corresponding inverse h^{ij}

$$h^{ij}(q_1, q_2) = \begin{pmatrix} \frac{(4+q_1^2+q_2^2)^2}{16} & 0 \\ 0 & \frac{(4+q_1^2+q_2^2)^2}{16} \end{pmatrix} \tag{1.33}$$

Metrics that differ only by a multiplicative factor are conformal equivalent. The stereographic projection endows the plane with a metric conformal to the regular Euclidean metric.

Chapter 2
Calibration of Omnidirectional Cameras Using a DLT-Like Approach

Abstract In this chapter, we present a new calibration technique that is valid for all single-viewpoint catadioptric cameras. We are able to represent the projection of 3D points on a catadioptric image linearly with a 6×10 projection matrix, which uses *lifted coordinates* for image and 3D points. This projection matrix can be linearly computed from 3D to 2D correspondences (minimum 20 points distributed in three different planes). We show how to decompose it to obtain intrinsic and extrinsic parameters. Moreover, we use this parameter estimation followed by a nonlinear optimization to calibrate various types of cameras. Our results are based on the sphere camera model. We test our method both with simulations and real images, and we analyze the results performing a 3D reconstruction from two omnidirectional images.

2.1 Introduction

Since their introduction to the computer vision community, catadioptric omnidirectional cameras have been utilized in many application areas such as surveillance, tracking, tele-presence, visual navigation, localization and SLAM, structure from motion, active vision, visual odometry, photogrammetry, camera networks, reconstruction of cultural heritage, among others.

Camera calibration is essential when we want to extract metric information from images. It establishes a relationship between the 3D rays and their corresponding pixels in the image. This relationship makes possible to measure distances in a real world from their projections on the images (Faugeras 1993). Camera calibration is basically composed of two steps. The first step consists of modeling the physical and optical behavior of the sensor through a geometric-mathematical model. There exist several approaches that propose different models to deal with central catadioptric systems (Kang 2000; Svoboda and Pajdla 2002; Scaramuzza et al. 2006; Toepfer and Ehlgen 2007; Geyer and Daniilidis 2000). The second step consists of estimating

L. Puig and J. J. Guerrero, *Omnidirectional Vision Systems*, SpringerBriefs in Computer Science, DOI: 10.1007/978-1-4471-4947-7_2, © Luis Puig 2013

the parameters that compose this model using direct or iterative methods. These parameters are of two types, intrinsic and extrinsic. The intrinsic parameters basically consider how the light is projected through the mirror and the lens onto the image plane of the sensor. The extrinsic parameters describe the position and orientation of the catadioptric system with respect to a world coordinate system.

Several methods have been proposed for calibration of catadioptric systems. Some of them consider estimating the parameters of the parabolic (Geyer and Danilidis 2002a; Kang 2000), hyperbolic (Orghidan et al. 2003), and conical (Cauchois et al. 1999) mirrors together with the camera parameters. Some others separate the geometry of the mirror from the calibration of the conventional camera (Svoboda and Pajdla 2002; Moral and Fofi 2007). Calibration of outgoing rays based on a radial distortion model is another approach. Kannala and Brandt (2004) used this approach to calibrate fisheye cameras. Scaramuzza et al. (2006) and Tardit et al. (2006) extended the approach to include central catadioptric cameras as well. Mei and Rives (2007), on the other hand, developed another Matlab calibration toolbox that estimates the parameters of the sphere camera model. Parameter initialization is done by user input, namely, the location of the principal point and depiction of a real-world straight line in the omnidirectional image (for focal length estimation).

In this chapter a new method to calibrate central catadioptric systems is proposed. This method was previously shown in Bastanlar et al. (2008) and an improved version was presented in Puig et al. (2011), which also includes: the studying of the use only two planes and additional constraints to perform the calibration; the relation between the intrinsic parameters of the sphere camera model and the actual camera; and a 3D reconstruction experiment to show the effectiveness of the approach. In this work, the calibration theory of central cameras proposed by Sturm and Barreto (2008) is put into practice. We compute the generic projection matrix, P_{cata}, with 3D–2D correspondences, using a straightforward DLT-like [Direct Linear Transform (Abdel-Aziz and karara 1971)] approach, i.e., by solving a linear equation system. Then, we decompose P_{cata} to estimate intrinsic and extrinsic parameters. Having these estimates as initial values of system parameters, we optimize the parameters based on minimizing the reprojection error. A software version of our method is available at the webpage.[1] When compared with alternate techniques our approach has the advantage of not requiring input for parameter initialization and being able to calibrate perspective cameras as well. Although it only requires a single catadioptric image, it must be of a 3D calibration object.

2.2 Generic Projection Matrix P_{cata}

As explained in Sect. 1.2, a 3D point is mathematically projected to two image points. Sturm and Barreto (2008) represented these two 2D points via the degenerate dual conic generated by them, i.e., the dual conic containing exactly the lines going

---

[1] http://webdiis.unizar.es/~lpuig/DLTOmniCalibration/Toolbox.tar.gz

through at least one of the two points. Let the two image points be $\mathbf{q}_+$ and $\mathbf{q}_-$; the dual conic is then given by

$$\Omega \sim \mathbf{q}_+\mathbf{q}_-^\mathsf{T} + \mathbf{q}_-\mathbf{q}_+^\mathsf{T} \tag{2.1}$$

The vectorized matrix of the conic can be computed as shown below using the lifted 3D point coordinates, intrinsic and extrinsic parameters.

$$\mathbf{v}_{sym}(\Omega) \sim \widehat{\mathsf{K}}_{6\times6}\mathsf{X}_\xi\widehat{\mathsf{R}}_{6\times6}\,(\mathsf{I}_6\ \mathsf{T}_{6\times4})\,\hat{\mathbf{Q}}_{10} \tag{2.2}$$

Here, R represents the rotation of the catadioptric camera. X_ξ and $\mathsf{T}_{6\times4}$ depend only on the sphere model parameter ξ and the position of the catadioptric camera $\mathbf{C} = (t_x, t_y, t_z)$ respectively, as shown here:

$$\mathsf{X}_\xi = \begin{pmatrix} 1 & 0 & 0 & 0 & 0 & 0 \\ 0 & 1 & 0 & 0 & 0 & 0 \\ 0 & 0 & 1 & 0 & 0 & 0 \\ 0 & 0 & 0 & 1 & 0 & 0 \\ 0 & 0 & 0 & 0 & 1 & 0 \\ -\xi^2 & 0 & -\xi^2 & 0 & 0 & 1-\xi^2 \end{pmatrix} \tag{2.3}$$

$$\mathsf{T}_{6\times4} = \begin{pmatrix} -2t_x & 0 & 0 & t_x^2 \\ -t_y & -t_x & 0 & t_xt_y \\ 0 & -2t_y & 0 & t_y^2 \\ -t_z & 0 & -t_x & t_xt_z \\ 0 & -t_z & -t_y & t_yt_z \\ 0 & 0 & -2t_z & t_z^2 \end{pmatrix} \tag{2.4}$$

Thus, a 6×10 **catadioptric projection matrix**, P_{cata}, can be expressed by its intrinsic and extrinsic parameters, like the projection matrix of a perspective camera.

$$\mathsf{P}_{cata} = \underbrace{\widehat{\mathsf{K}}\mathsf{X}_\xi}_{\mathsf{A}_{cata}}\ \underbrace{\widehat{\mathsf{R}}_{6\times6}(\mathsf{I}_6\mathsf{T}_{6\times4})}_{\mathsf{T}_{cata}} \tag{2.5}$$

2.2.1 Computation of the Generic Projection Matrix

Here, we show the way used to compose the equations using 3D–2D correspondences to compute P_{cata}. Analogous to the perspective case ($[\mathbf{q}]_\times\mathsf{PQ} = \mathbf{0}$), we write the constraint based on the lifted coordinates (Sturm and Batteto 2008):

$$\widehat{[\mathbf{q}]}_\times\,\mathsf{P}_{cata}\,\hat{\mathbf{Q}} = \mathbf{0} \tag{2.6}$$

This is a set of 6 linear homogeneous equations in the coefficients of P_{cata}. Using the Kronecker product, this can be written in terms of the 60-vector $\mathbf{p}_{\text{cata}}$ containing the 60 coefficients of P_{cata}:

$$(\hat{\mathbf{Q}}^{\mathsf{T}} \otimes \widehat{[\mathbf{q}]}_{\times})\mathbf{p}_{\text{cata}} = \mathbf{0}_6 \tag{2.7}$$

Stacking these equations for n 3D–2D correspondences gives a system of equations of size $6n \times 60$, which can be solved by linear least squares, e.g., using the Singular Value Decomposition (SVD). Note that the minimum number of required correspondences is 20: a 3×3 skew symmetric matrix has rank 2, its lifted counterpart rank 3. Therefore, each correspondence provides only 3 independent linear constraints.

2.3 Generic Projection Matrix and Calibration

The calibration process consists of getting the intrinsic and extrinsic parameters of a camera. Once P_{cata} has been computed from point correspondences, our purpose is to decompose P_{cata} as in (4.10). Consider first the leftmost 6×6 submatrix of P_{cata}:

$$\mathsf{P}_s \sim \widehat{\mathsf{K}} \mathsf{X}_\xi \widehat{\mathsf{R}} \tag{2.8}$$

Let us define $\mathsf{M} = \mathsf{P}_s \mathsf{D}^{-1} \mathsf{P}_s^{\mathsf{T}}$. Using the properties given in (1.24) and knowing that for a rotation matrix $\mathsf{R}^{-1} = \mathsf{R}^{\mathsf{T}}$, we can write $\widehat{\mathsf{R}}^{-1} = \mathsf{D}^{-1}\widehat{\mathsf{R}}^{\mathsf{T}}\mathsf{D}$. And from that we obtain $\mathsf{D}^{-1} = \widehat{\mathsf{R}}\mathsf{D}^{-1}\widehat{\mathsf{R}}^{\mathsf{T}}$ which we use to eliminate the rotation parameters:

$$\mathsf{M} \sim \widehat{\mathsf{K}} \mathsf{X}_\xi \widehat{\mathsf{R}} \mathsf{D}^{-1}\widehat{\mathsf{R}}^{\mathsf{T}} \mathsf{X}_\xi^{\mathsf{T}} \widehat{\mathsf{K}}^{\mathsf{T}} = \widehat{\mathsf{K}} \mathsf{X}_\xi \, \mathsf{D}^{-1} \mathsf{X}_\xi^{\mathsf{T}} \widehat{\mathsf{K}}^{\mathsf{T}} \tag{2.9}$$

Equation (2.9) holds up to scale, i.e., there is a λ with $\mathsf{M} = \lambda \widehat{\mathsf{K}} \mathsf{X}_\xi \, \mathsf{D}^{-1} \mathsf{X}_\xi^{\mathsf{T}} \widehat{\mathsf{K}}^{\mathsf{T}}$. For initialization we assume that the camera is well aligned with the mirror axis, i.e., assume that $\mathsf{R}_p = \mathsf{I}$, thus $\mathsf{K} = \mathsf{A}_p = \begin{pmatrix} f & 0 & c_x \\ 0 & f & c_y \\ 0 & 0 & 1 \end{pmatrix}$.

We then use some elements of M to extract the intrinsic parameters:

$$\begin{aligned}
M_{16} &= \lambda \left(-(f^2\xi^2) + c_x^2 \left(\xi^4 + c_x(1 - \xi^2)^2\right)\right) \\
M_{44} &= \lambda \left(\frac{f^2}{2} + c_x^2 \left(2\xi^4 + (1 - \xi^2)^2\right)\right) \\
M_{46} &= \lambda c_x \left(2\xi^4 + (1 - \xi^2)^2\right) \\
M_{56} &= \lambda c_y \left(2\xi^4 + (1 - \xi^2)^2\right) \\
M_{66} &= \lambda \left(2\xi^4 + (1 - \xi^2)^2\right)
\end{aligned} \tag{2.10}$$

The intrinsic parameters are computed as follows:

$$c_x = \frac{M_{46}}{M_{66}} \qquad c_y = \frac{M_{56}}{M_{66}} \qquad \xi = \sqrt{\frac{\frac{M_{16}}{M_{66}} - c_x^2}{-2\left(\frac{M_{44}}{M_{66}} - c_x^2\right)}}$$

$$f = \sqrt{2(2\xi^4 + (1 - \xi^2)^2)\left(\frac{M_{44}}{M_{66}} - c_x^2\right)} \tag{2.11}$$

After extracting the intrinsic parameters matrix A_{cata} of the projection matrix, we are able to obtain the 6×10 extrinsic parameters matrix T_{cata} by multiplying P_{cata} with the inverse of A_{cata}:

$$T_{cata} = \widehat{R}_{6\times6}(I_6 T_{6\times4}) \sim (\widehat{K}X_\xi)^{-1} P_{cata} \tag{2.12}$$

Hence, the leftmost 6×6 part of T_{cata} will be the estimate of the lifted rotation matrix $\widehat{R}_{est}$. If we multiply the inverse of this matrix with the rightmost 6×4 part of T_{cata}, we obtain an estimate for the translation ($T_{6\times4}$). This translation should have an ideal form as given in (2.4) and we are able to identify translation vector elements (t_x, t_y, t_z) from it straightforwardly.

We finally have to handle the fact that the estimated $\widehat{R}_{est}$ will not, in general, be an exact lifted rotation matrix. This lifted rotation matrix in particular is oversized since it considers the lifting of a full rotation matrix $\widehat{R} = \widehat{R}_z(\gamma)\widehat{R}_y(\beta)\widehat{R}_x(\alpha)$. For illustration in (2.13) we show the lifting of a rotation matrix around the x-axis.

$$\widehat{R}_x(\alpha) = \begin{pmatrix} 1 & 0 & 0 & 0 & 0 & 0 \\ 0 & \cos\alpha & 0 & -\sin\alpha & 0 & 0 \\ 0 & 0 & \cos^2\alpha & 0 & -2\cos\alpha\sin\alpha & \sin^2\alpha \\ 0 & \sin\alpha & 0 & \cos\alpha & 0 & 0 \\ 0 & 0 & \cos\alpha\sin\alpha & 0 & \cos^2\alpha - \sin^2\alpha & -\cos\alpha\sin\alpha \\ 0 & 0 & \sin^2\alpha & 0 & 2\cos\alpha\sin\alpha & \cos^2\alpha \end{pmatrix} \tag{2.13}$$

Since P_{cata} has been estimated up to scale it is impossible to extract the rotation components from single elements of $\widehat{R}_{est}$. To deal with this problem we algebraically manipulate the ratios between the elements of this lifted matrix and we extract the angles one by one. First, we recover the rotation angle around the z axis, $\gamma = \tan^{-1}\left(\frac{\widehat{R}_{est,51}}{\widehat{R}_{est,41}}\right)$. Then, $\widehat{R}_{est}$ is modified by being multiplied by the inverse of the rotation around the z axis, $\widehat{R}_{est} = \widehat{R}_z^{-1}(\gamma)\widehat{R}_{est}$. Then, the rotation angle around the y

axis, β, is estimated and $\widehat{R}_{est}$ is modified $\beta = \tan^{-1}\left(\frac{-\widehat{R}_{est,52}}{\widehat{R}_{est,22}}\right)$, $\widehat{R}_{est} = \widehat{R}_y^{-1}(\beta)\,\widehat{R}_{est}$.

Finally, the rotation angle around the x axis, α, is estimated as $\alpha = \tan^{-1}\left(\frac{\widehat{R}_{est,42}}{\widehat{R}_{est,22}}\right)$.

2.3.1 Other Parameters of Nonlinear Calibration

The intrinsic and extrinsic parameters extracted in closed-form in Sect. 2.3 are not always adequate to model a real camera. Extra parameters are needed to correctly model the catadioptric system, namely, tilting and lens distortions.

As mentioned before $\widehat{K} = \widehat{A_p R_p} = \widehat{A}_p \widehat{R}_p$ where R_p is the rotation between camera and mirror coordinate systems, i.e., tilting. Tilting has only R_x and R_y components, because rotation around the optical axis, R_z, is coupled with the external rotation around the z axis of the entire catadioptric system. Note that tilting angles of the sphere camera model are not equivalent to the tilting angles of the actual perspective camera looking at the mirror.

As is well known, imperfections due to lenses are modeled as distortions for camera calibration. Radial distortion models contraction or expansion with respect to the image center and tangential distortion models lateral effects. To add these distortion effects to our calibration algorithm, we employed the approach of Heikkila and Silven (1997).

Radial distortion:

$$\begin{aligned}
\Delta x &= x(k_1 r^2 + k_2 r^4 + k_3 r^6 + \cdots) \\
\Delta y &= y(k_1 r^2 + k_2 r^4 + k_3 r^6 + \cdots)
\end{aligned} \tag{2.14}$$

where $r = \sqrt{x^2 + y^2}$ and $k_1, k_2 \ldots$ are the radial distortion parameters. We observe that estimating two parameters is enough for an adequate estimation. Tangential distortion:

$$\begin{aligned}
\Delta x &= 2p_1 xy + p_2(r^2 + 2x^2) \\
\Delta y &= p_1(r^2 + 2y^2) + 2p_2 xy
\end{aligned} \tag{2.15}$$

where $r = \sqrt{x^2 + y^2}$ and p_1, p_2 are the tangential distortion parameters.

Once we have identified all the parameters to be estimated we perform a nonlinear optimization to compute the whole model. We use the Levenberg–Marquardt method (LM).[2] The minimization criterion is the root mean square (RMS) of distance between a measured image point and its reprojected correspondence. Since the projection equations we use map 3D points to dual image conics, we have to extract the two potential image points from it. The one closer to the measured point is selected

---

[2] Method provided by the function **lsqnonlin** in Matlab.

and then the reprojection error measured. We take as initial values the parameters obtained from P_{cata} and initialize the additional 4 distortion parameters and the tilt angles in R_p, by zero.

2.3.2 Algorithm to Compute P_{cata}

Here we summarize the algorithm used to compute the generic projection matrix P_{cata}.

1. **Linear Solution**. Using 3D–2D correspondences we compute P_{cata} by a DLT-like approach.
2. **Intrinsic/Extrinsic Parameter Extraction**. Assuming that the perspective camera is perfectly aligned with the mirror axis, i.e., there is no tilting and that the images are not distorted. We extract from the linear solution, the intrinsic (ξ, f, c_x, c_y) and extrinsic ($\alpha, \beta, \gamma, t_x, t_y, t_z$) parameters in closed-form.
3. **Initialization Vector**. An initialization vector is constructed with the extracted parameters. Two parameters are added to consider the tilting angles (r_x, r_y) and four more corresponding to the radial (k_1, k_2) and tangential (p_1, p_2) distortion.
4. **Nonlinear Optimization Process**. Using this vector as an initialization vector, we perform a nonlinear optimization process using the LM algorithm. The minimization criterion is the reprojection error.

2.4 Theoretical and Practical Issues

In the last section, we explained that twenty 3D–2D correspondences are enough to compute the calibration of the central catadioptric systems. In principle these twenty correspondences can be located anywhere inside the FOV of the catadioptric system. Since we want to construct a feasible calibration system based on planar patterns we restrict the 3D points to be located in planes. From simulations we observed that the minimum number of planes where the 3D points should be located is three in the general case. In particular, two planes can be used to compute P_{cata} if several constraints are imposed, but the simplicity of using linear equations is lost.

Since we restrict the calibration points to lie on planes (planar grid-based calibration) some degeneracies can appear if the calibration points are located in a particular configuration. Something similar to the pin-hole camera case with the twisted cubic (Buchanan 1988), for which calibration fails even if the points lie on more than two planes. However, a complete analysis of such degeneracies is out of the scope of this book.

In this section, we present a proof that points lying in three different planes are required to linearly and uniquely compute the generic projection matrix P_{cata}. We also show that under several assumptions we can compute P_{cata} from points lying in just two planes.

2.4.1 Three Planes are Needed to Compute P_{cata} Using Linear Equations

Here we show that in order to compute P_{cata}, the 3D calibration points must lie in at least 3 different planes. We first prove that two planes are not sufficient. Let Π_1 and Π_2 be the two planes. Hence, each calibration point Q satisfies $\left(\Pi_1^T Q\right)\left(\Pi_2^T Q\right) = 0$. This can be written as a linear constraint on the lifted calibration points: $p^T \hat{Q} = 0$, where the 10-vector p depends exactly on the two planes. Thus, if P_{cata} is the true 6×10 projection matrix, then adding some multiple of p^T to any row of P_{cata} gives another 6×10 projection matrix, $\bar{P}_{cata}$, which maps the calibration points to the same image entities as the true projection matrix. We may write the ambiguity as

$$\bar{P}_{cata} = P_{cata} + v p^T \tag{2.16}$$

where v is a 6-vector and represents the six degrees of freedom (DoF) on P_{cata} that can not be recovered using only linear projection equations and calibration points located in only two planes. This is not the case for perspective cameras, where two planes are enough to compute the 3×4 perspective projection matrix.

For three planes, there is no linear equation as above that holds for all calibration points. Hence, also supported by our experiments, it seems plausible that three planes are sufficient for uniquely computing the projection matrix. Note that by planes we do not mean that calibration grids have to be composed of three or more planar grids. The planes can be virtual: whenever it is possible to fit the two planes to the whole set of 3D points, P_{cata} can not be computed.

2.4.2 Adding Constraints to Estimate the Projection Matrix from Points on Two Planes Only

In the last section we observe that to compute P_{cata} linearly and uniquely, 3D points must be sufficiently well distributed, such that no two planes contain all of them. In this section, we analyze what prior information allows nevertheless to compute the calibration parameters using two planes. We know by (2.16) that the true projection matrix is related to any other solution by

$$P_{cata} = \bar{P}_{cata} - v p^T \tag{2.17}$$

Consider the equation to eliminate the extrinsic parameters:

$$M \sim P_s D^{-1} P_s^T \tag{2.18}$$

where P_s is the leftmost 6×6 submatrix of P_{cata}. Now we redefine it as follows:

$$M \sim (\bar{P}_s - vp_s^T)D^{-1}(\bar{P}_s - vp_s^T)^T \tag{2.19}$$

where $\bar{P}_s$ is the leftmost 6×6 submatrix of $\bar{P}_{\text{cata}}$ and p_s is the first 6 elements of the 10-vector p. Assuming that the two planes are perpendicular to each other, we can write $\Pi_1 = [1, 0, 0, 0]^T$ and $\Pi_2 = [0, 1, 0, 0]^T$ which gives us $p_s = [0, 1, 0, 0, 0, 0]^T$ (we obtain p by $v_{\text{sym}}(\Pi_1\Pi_2^T + \Pi_2\Pi_1^T)$ since $\Pi_1\Pi_2^T$ represents a degenerate dual conic on which all Q lie).

Let us develop (2.19):

$$M \sim \underbrace{\bar{P}_s D^{-1}\bar{P}_s^T}_{\tilde{M}} - \underbrace{\bar{P}_s D^{-1}p_s}_{b} v^T - v\underbrace{p_s^T D^{-1}\bar{P}_s}_{b^T} + v\underbrace{p_s^T D^{-1}p_s}_{\rho} v^T \tag{2.20}$$

$$M \sim \tilde{M} - bv^T - vb^T + \rho vv^T \tag{2.21}$$

We can compute ρ, it is $\frac{1}{2}(D_{22} = 2)$. So we just need to obtain elements of v to recover P_{cata}. The principal point can be computed using different approaches, one of these is shown in Mei and Rives (2007), which requires the user interaction. Let us suppose we know the principal point (c_x, c_y), and we put the origin of the image reference system on it ($c_x = 0$, $c_y = 0$). Then we have:

$$M = \begin{pmatrix} f^4 & 0 & 0 & 0 & 0 & -f^2\xi^2 \\ 0 & \frac{f^4}{2} & 0 & 0 & 0 & 0 \\ 0 & 0 & f^4 & 0 & 0 & -f^2\xi^2 \\ 0 & 0 & 0 & \frac{f^2}{2} & 0 & 0 \\ 0 & 0 & 0 & 0 & \frac{f^2}{2} & 0 \\ -f^2\xi^2 & 0 & -f^2\xi^2 & 0 & 0 & 2\xi^4 + (1-\xi^2)^2 \end{pmatrix} \tag{2.22}$$

From this matrix we can extract 6 equations to solve for the elements of v. For example: $M_{11} - M_{33} = 0$, $M_{11} - 2M_{22} = 0$, $M_{44} - M_{55} = 0$, $M_{13} = 0$, $M_{35} = 0$, $M_{56} = 0$.

We test the case where $f_x = f_y$ using simulated data with perfect 3D–2D correspondences. We observe that as explained in theory, the only modified column is the second one, described by the vector $p_s = [0, 1, 0, 0, 0, 0]^T$. In this case we are able to obtain the correct P_{cata}. However, when we added Gaussian noise to the 3D–2D correspondences, more than one column is modified making very difficult to recover the real projection matrix. Therefore, we conclude that the approach using points lying in just two planes is not suitable to compute the generic projection matrix in real situations. We continue our experiments with calibration grids having three planes.

2.5 Calibration Experiments with a Simulated Environment

We use a simulated calibration object having 3 planar faces which are perpendicular to each other. The size of a face is 50×50 cm. There are a total of 363 points, since each face has 11×11 points and the distance between points is 5 cm. The omnidirectional image fits in a 1 Megapixel square image. To represent the real world points we expressed the coordinates in meters, so they are normalized in a sense. This is important because we observed that using large numerical values causes bad estimations with noisy data in the DLT algorithm. Normalization of image coordinates is also performed since we observed a positive effect both on estimation accuracy and the convergence time. Therefore, in the presented experiments, 3D point coordinates are in meters and image coordinates are normalized to be in the same order of magnitude, this is performed by dividing the image coordinates by a constant.

We performed experiments for different settings of intrinsic parameters and varying position of the 3D calibration grid. We especially tested the accuracy of calibration to variations in the intrinsic parameters (ξ and f), the distance between the camera and the grid and the orientation of the grid w.r.t. the camera. In all these cases, we measure the errors in final estimates of ξ and f, the main parameters of the sphere camera model. Errors are depicted in Fig. 2.1, where an individual graph is plotted for each case for clarity. In all experiments, Gaussian noise with $\sigma = 1$ pixel is added to the actual coordinates of grid corners. The plotted errors are $\mathrm{err}_\xi = 100 \cdot |\xi_{\mathrm{nonlin}} - \xi_{\mathrm{real}}| / \xi_{\mathrm{real}}$ and $\mathrm{err}_f = 100 \cdot |f_{\mathrm{nonlin}} - f_{\mathrm{real}}| / f_{\mathrm{real}}$. For all the nodes in the graphs, the experiment was repeated 100 times and the mean value of estimates is plotted.

Figure 2.1a shows the effect of changing distance between the camera and the grid. From left to right in the graph distance-to-grid increases and distance values

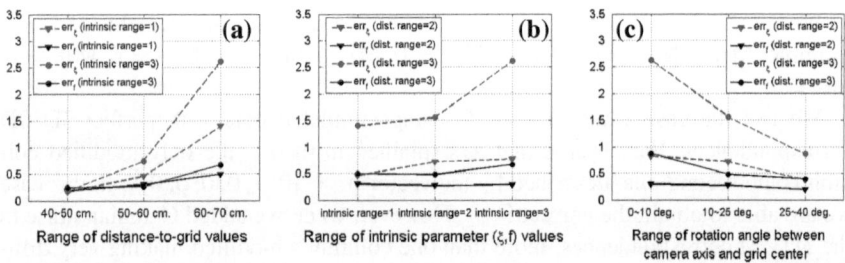

Fig. 2.1 Relative errors for ξ and f after nonlinear optimization (in percent) for varying intrinsic parameters and varying position of the 3D calibration grid. For all the nodes in the graphs, the experiment was repeated 100 times and the mean value of estimates is plotted. Real intrinsics, distance and orientation values are selected randomly from the ranges given in x-axis. Intrinsic parameters range 1: $(\xi, f) = [(0.96, 360) (0.84, 300)]$, range 2: $(\xi, f) = [(0.84, 300) (0.72, 250)]$, range 3: $(\xi, f) = [(0.72, 250) (0.60, 210)]$. Distance-to-grid (in cm) range 1: [40 50], range 2: [50 60], range 3: [60 70]. In **a**, **b** and **c**, errors depicted versus increasing distance-to-grid, decreasing (ξ, f) pairs and increasing rotation angle respectively

are selected randomly within the given ranges. When the distance is small, we reach an "optimal" position, such that the grid fills the image well. As the grid moves away from the omnidirectional camera, its image gets smaller and smaller. Examples of the omnidirectional images generated are shown in Fig. 2.2. In Fig. 2.2a, distance-to-grid is 45 cm, whereas in Fig. 2.2b it is 60 cm. The quality of parameter estimation decreases with increasing distance. Since the grid covers a smaller area, the same amount of noise (in pixels) affects the nonlinear optimization more and errors in nonlinear results increase as can be expected. We observe the importance of a good placement of the calibration grid, i.e., such that it fills the image as much as possible.

Figure 2.1b shows the effect of real ξ and f values on the estimation error (for two different distance-to-grid value ranges). From left to right in the graph, ξ and f values decrease. They decrease in parallel, otherwise decreasing ξ with fixed f would cause grid to get smaller in the image. We truncated (ξ, f) pairs at $\xi = 0.6$ since even smaller ξ values are unlikely for omnidirectional cameras. We observe that larger (ξ, f) values produce slightly better results especially for increased distances. This observation can also be made in Fig. 2.1a since the errors are depicted with two different ranges of intrinsic parameter values. The reason is that for fixed distance-to-grid values, higher (ξ, f) spreads the grid points to a larger area in the image, which decreases the effect of noise. Observe Fig. 2.2b with Fig. 2.2c, where distance-to-grid values are equal but Fig. 2.2b has higher (ξ, f).

Figure 2.1c shows the effect of changing orientation of the grid w.r.t. the camera. This is expressed in terms of the angle between the optical axis of the omnidirectional camera and the grid center. The grid is not rotated independently from the camera axis because camera (mirror) has to see the inside of the 3D grid always. Figure 2.2d shows the case when the grid is rotated so that the angle between its center and camera optical axis is 40°. Compared with Fig. 2.2b, where the intersection of the three planes of the grid is at the image center. We observe improvement with rotation specially for increased distance-to-grid since grid points are more spread and effect of noise decreases.

In Table 2.1, we list the results of the algorithm after linear (DLT) and nonlinear steps for a few cases. Our main observation is that the errors in linear estimates, ξ_{DLT} and f_{DLT}, are biased (values are smaller than they should be). For all the cases, however, the true intrinsic parameters are reached after nonlinear optimization, modulo errors due to noise.

2.5.1 Estimation Errors for Different Camera Types

Here we discuss the intrinsic and extrinsic parameter estimation for the two most common catadioptric systems: hypercatadioptric and paracatadioptric, with hyperbolic and parabolic mirrors respectively. We also discuss calibration results for perspective cameras.

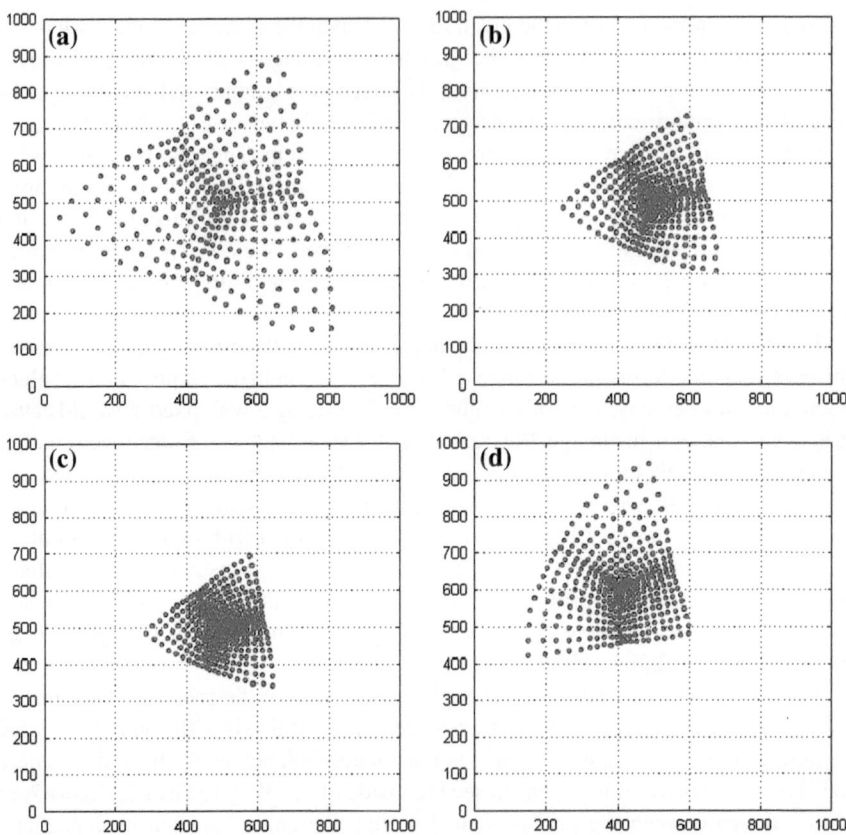

Fig. 2.2 Omnidirectional images generated with varying intrinsics, distance-to-grid, and orientation. **a** $(\xi, f) = (0.96, 360)$, distance $= 45$ cm, no rotation. **b** $(\xi, f) = (0.96, 360)$, distance $= 60$ cm, no rotation. **c** $(\xi, f) = (0.76, 270)$, distance $= 60$ cm, no rotation. **d** $(\xi, f) = (0.96, 360)$, distance $= 60$ cm, rotated by $40°$

2.5.1.1 Hypercatadioptric System

Table 2.2 shows nonlinear optimization results including the rotation and translation parameters for fixed intrinsic parameters which corresponds to a hypercatadioptric system. 3D pattern is used at the "optimal" grid position, i.e., it fills the omnidirectional image like Fig. 2.2a. Results are in accordance with Table 2.1 and Fig. 2.1.

2.5.1.2 Paracatadioptric System

Here $\xi = 1$, which has a potential to disturb the estimations because X_ξ becomes a singular matrix. We observe that the results of the DLT algorithm are not as close to the real values when compared to the hypercatadioptric system (cf. initial values

Table 2.1 Initial and optimized estimates with different intrinsics and distance-to-grid values

	Distance-to-grid			
	45 cm		60 cm	
ξ_{real}	0.96	0.8	0.96	0.80
f_{real}	360	270	360	270
ξ_{DLT}	0.54	0.40	0.04	0.03
f_{DLT}	361	268	243	190
ξ_{nonlin}	0.96	0.80	0.98	0.78
f_{nonlin}	360	270	365	266
err_{ξ}	0.0	0.0	2.1	2.5
err_f	0.0	0.1	1.4	1.5

Amount of noise: $\sigma = 1$ pixel. ξ_{DLT}, f_{DLT} and ξ_{nonlin}, f_{nonlin} are the results of the DLT algorithm and nonlinear optimization respectively, err_{ξ} and err_f are the relative errors, in percent after nonlinear optimization

Table 2.2 Nonlinear optimization results for a hypercatadioptric system, 10 parameters (rotation, translation, and intrinsic) are optimized

	Real values	$\sigma = 0.5$		$\sigma = 1$	
		Initial	Estimated	Initial	Estimated
f	360	361	360	354	360
c_x	500	503	500	505	500
c_y	500	498	500	509	500
ξ	0.96	0.84	0.96	0.53	0.96
$R_x(\alpha)$	−0.62	−0.60	−0.62	−0.40	−0.62
$R_y(\beta)$	0.62	0.62	0.62	0.65	0.62
$R_z(\gamma)$	0.17	0.15	0.17	0.18	0.17
t_x	0.30	0.38	0.30	0.45	0.30
t_y	0.30	0.40	0.30	0.44	0.30
t_z	0.20	0.05	0.20	0.01	0.20
RMSE	–	–	0.70	–	1.42

Distance-to-grid is 45 cm and grid center coincides with camera optical axis (no rotation)

in Table 2.2). However, the nonlinear optimization estimates the parameters as successful as the hypercatadioptric examples given in Table 2.2.

2.5.1.3 Perspective Camera

In the sphere camera model, $\xi = 0$ corresponds to the perspective camera. Our estimations in linear and nonlinear steps are as successful as with the hypercatadioptric case and thus not shown in detail here.

2.5.2 Tilting and Distortion

It seems intuitive that small amounts of tangential distortion and tilting have a similar effect on the image. In our simulations we observed that trying to estimate both of them does not succeed. Therefore, we investigate if we can estimate tangential distortion of camera optics by tilt parameters, or estimate tilt in the system by tangential distortion parameters.

When there exists no tilt but tangential distortion and we try to estimate tilting parameters, we observed that the direction and amount of $tilt_x$, $tilt_y$, c_x and c_y changes proportionally to the tangential distortion applied and the RMSE decreases. However, the RMSE does not reach as low values as when there is no distortion. In the noiseless case, for example, the RMSE is not zero. Hence, we concluded that tilt parameters compensate the tangential distortion effect up to some extent, but not perfectly. We also investigated if tilting can be compensated by tangential distortion parameters and we had very similar results. Thus, tangential distortion parameters have the same capability to estimate tilting.

2.6 Experiments with Real Images Using a 3D Pattern

In this section we perform experiments of camera calibration using a 3D pattern, cf. Fig. 2.3a. The 3D pattern has been measured accurately doing a photogrammetric reconstruction by bundle adjustment. We use 6 convergent views taken with a calibrated high-resolution camera (Canon EOS 5D with 12.8 Megapixel) and software PhotoModeler. The estimated accuracy of the 3D model is better than 0.1 mm. The omnidirectional images were acquired using a catadioptric system with a hyperbolic

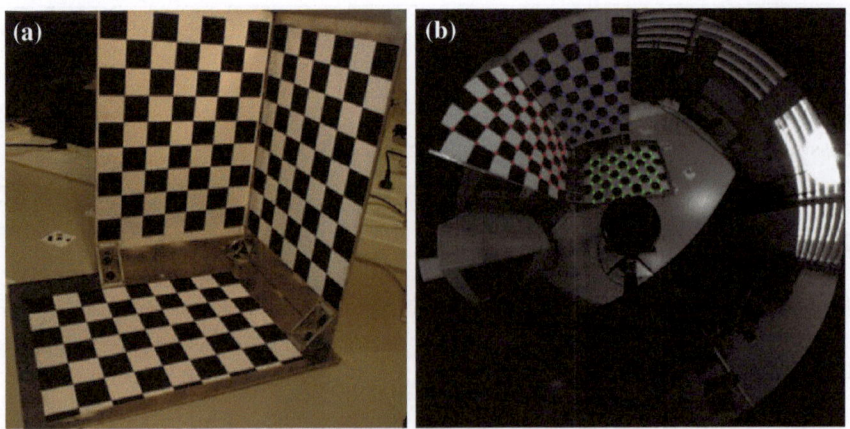

Fig. 2.3 **a** 3D pattern, **b** Omnidirectional image of the 3D pattern (1024 × 768 pixels)

Table 2.3 Parameters estimated using either tangential distortion or tilting angles

	Real	Using distortion	Using tilting
f	279.84	297.24	306.11
c_x	531.83	528.08	552.75
c_y	407.98	406.28	427.89
ξ	0.96	0.86	0.93
RMSE	0	0.34	0.27

mirror.[3] We computed the projection matrix P_{cata} from a total of 144 3D to 2D correspondences and extracted the intrinsic and extrinsic parameters as explained in Sect. 2.2. From simulations, we observed that we have better and faster estimations if the 3D–2D correspondences are in the same order of magnitude. So 3D points are given in meters and 2D points are normalized in all the experiments. A second evaluation of the calibration accuracy is performed by a Structure from Motion experiment from two omnidirectional images.

2.6.1 Intrinsic Parameters

The first experiment is focused on obtaining the intrinsic parameters from P_{cata} to get initial estimates of these values. As mentioned previously, we do not compute tilting and distortion parameters from P_{cata} but it is possible to include them in the nonlinear optimization. From simulations we observed that we can compute either the tangential distortion or the tilting parameters which are coupled and cannot be separated. We tested which one of these (tangential distortion and tilting) can deal better with the intrinsic parameter estimation. Table 2.3 shows a comparison of the estimations performed with these two options. The real values given in the table were computed using the calibration data of the perspective camera (previously calibrated) and the mirror parameters (provided by the manufacturer).

Catadioptric camera calibration using tilting gives a better RMSE but the intrinsic values obtained are far from the real ones. Estimation using distortion parameters increase slightly the RMSE but the intrinsic parameters are close to the real ones, except for ξ but this error can be attached to the configuration of the system (the optical center of the perspective camera may not be exactly located at the other focal point of the hyperbola describing the mirror) and not to the model.

After these results, we decided to use tangential distortion because it gives better results and depicts better the real catadioptric system.

In order to verify our approach we compare our intrinsic parameter estimates to the ones obtained by Mei and Rives (2007) (Table 2.4). As we can see neither Mei's approach nor P_{cata} approach can estimate the theoretic f and ξ parameters, but they give a good estimation to c_x and c_y. Mei computes the initial values directly from the

---

[3] Neovision H3S with XCD-X710 SONY camera.

Table 2.4 Comparison between our method and Mei's

	Theoretic	P_{cata} approach	Mei and Rives (2007)
f	279.84	297.24	298.65
ξ	0.96	0.86	0.72
c_x	531.83	528.02	528.15
c_y	407.98	406.28	403.39

Table 2.5 Rotation and translation of the camera with respect to the 3D pattern

	Experiment 1		Experiment 2		Experiment 3	
	Real	Estimated	Real	Estimated	Real	Estimated
R_x	−0.01	−0.02	−0.01	−0.003	−0.01	−0.002
R_y	0.02	0.02	0.02	0.01	0.02	0.03
R_z	–	–	–	–	–	–
t_x	0.39	0.39	0.39	0.39	0.39	0.38
t_y	0.21	0.21	0.33	0.33	0.23	0.23
t_z	−0.18	−0.18	−0.18	−0.18	−0.18	−0.18
RMSE	0.26		0.20		0.26	

Rotation angles are in radians. Translations are in meters. Real values were computed by the PhotoModeler software and a high-resolution camera

inner circle of the omnidirectional image and using information given by the user. Our approach computes all the initial values from P_{cata} in closed form.

2.6.2 Extrinsic Parameters

To obtain ground truth extrinsic parameters we have taken two additional images with the high-resolution camera, observing the omnidirectional camera and the pattern. These images are added to the ones used to measure the 3D pattern. From this set of images the orientation and translation of the camera with respect to the pattern are computed. Location of the focal point was difficult since the points are not easy to identify in the images and indeed inside the mirror.

We performed experiments with 3 different camera locations. Table 2.5 shows the rotations and translations obtained from these experiments. Using PhotoModeler software we were just able to compute the direction of the z-axis but not the rotation around it. So we just show rotation estimations for the x and y axis. We can observe that the extrinsic parameter estimation is performed with a good accuracy having an average error of 0.0096 radians for rotations and 0.0022 m for translations.

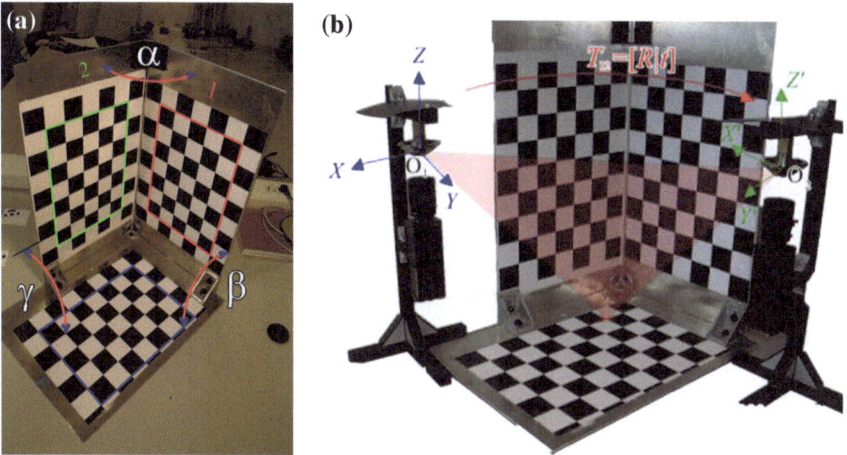

Fig. 2.4 **a** 3D pattern with the angles between the planes. **b** SfM configuration

2.6.3 Structure from Motion

The second experiment to evaluate the accuracy of our approach consists of obtaining the Structure and Motion (SfM) from two omnidirectional images observing the 3D pattern. Figure 2.4a shows the 3D pattern with the angles between the planes composing it. Figure 2.4b depicts the configuration used to perform the SfM experiment. Using the internal calibration provided by our method we compute the corresponding 3D rays from each omnidirectional image (Fig. 2.5). We use these correspondences of 3D rays to compute the essential matrix E which relates them. From this matrix we compute two projection matrices $P_1 = [I|0]$ and $P_2 = [R|t]$. Then, with these projection matrices and the 3D rays as input for a linear triangulation method (Hartley and Zisserman 2000) we compute an initial 3D reconstruction.

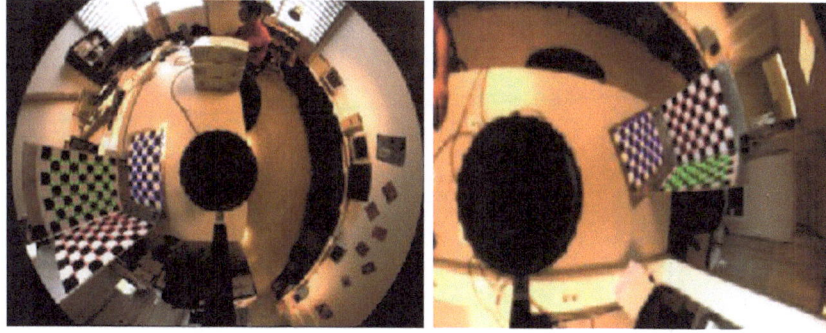

Fig. 2.5 Images used in the SfM experiment

Both the 3D reconstruction and the camera location are later refined by a nonlinear optimization process. We use 144 points which were extracted manually from the images. We measure the average error between the real 3D points and their estimations and the angle between the planes. We use as ground truth the data computed by the photogrammetric software. The angles between the planes as depicted in Fig. 2.4a are $\alpha = 90.06°$, $\beta = 89.60°$ and $\gamma = 90.54°$. The estimated values are $\alpha = 89.22°$, $\beta = 90.55°$ and $\gamma = 89.73°$. We have an average error of $0.86°$. We also measure the accuracy of the 3D points. The dimensions of the planar grids used in the 3D pattern are 210 mm $\times$ 294 mm. We compute the Euclidean distance between each reconstructed point and the ground truth. The average error is 1.03 mm.

2.7 Closure

In this chapter, we presented a new calibration technique based on the sphere camera model which is able to represent every single-viewpoint catadioptric system. We employed a generic 6×10 projection matrix, which uses *lifted coordinates* for image and 3D points. We estimated this projection matrix using 3D–2D correspondences. We use a single catadioptric image of a 3D calibration pattern. From the decomposition of this matrix we obtain an initial estimation of the intrinsic and extrinsic parameters of the catadioptric system. We used this parameter estimation as the initialization for a nonlinear optimization process. We are able to calibrate various types of cameras. This method was tested both with simulations and real images. Since the reprojection error is not definitive to show the good behavior of calibration approaches, we also present a Structure from Motion experiment to test the accuracy of our calibration method. For that reason we can provide error measurements in both pixels and millimeters.

Chapter 3
Comparison of Calibration Methods for Omnidirectional Cameras

Abstract The number of calibration methods for central catadioptric has increased in recent years. These methods are based on different camera models and they can either consider the central catadioptric system as a whole or as a separated camera and mirror system. Many times the user requires a versatile calibration solution without spending valuable time in the implementation of a particular method. In this chapter, we review the existing methods designed to calibrate any central omnivision system and analyze their advantages and drawbacks doing a deep comparison using simulated and real data. First we present a classification of such methods showing the most relevant characteristics of each particular method. Then we select the methods available as OpenSource and which do not require a complex pattern or scene. The evaluation protocol of calibration accuracy also considers 3D metric reconstruction combining omnidirectional images. Comparative results are shown and discussed in detail.

3.1 Introduction

Most of the applications of computer vision techniques, such as visual navigation, localization, and SLAM, structure from motion requires to recover metric information from the environment. This 3D information is crucial when the omnidirectional cameras interact in real scenarios. The metric information depends entirely on the complete calibration of the omnidirectional system. For these practical applications, the camera calibration is a basic step for subsequent and higher level tasks, and their final accuracy relies on the accuracy of the camera calibration. A considerable number of approaches to either calibrate central catadioptric systems or to calibrate fish-eye lens systems or both have been recently developed. Moreover, as we have observed they can use different projection models. With respect to central catadioptric systems, there exist some approaches that separate the calibration of the perspective/orthographic camera from the computation of the mirror parameters

L. Puig and J. J. Guerrero, *Omnidirectional Vision Systems*, SpringerBriefs in Computer Science, DOI: 10.1007/978-1-4471-4947-7_3, © Luis Puig 2013

(Svoboda and Pajdla 2002; Morel and Fofi 2007). However, most of them deal with the catadioptric system as a whole. Some of these methods use single or multiple views of a 2D pattern (Kannala and Brandt 2004; Scaramuzza et al. 2006; Mei and Rives 2007; Deng et al. 2007; Frank et al. 2007; Gasparini et al. 2009), 3D pattern (Puig et al. 2011; Bastanlar et al. 2008), cylinder pattern (Toepfer and Ehlgen 2007), some others use a single image containing features like lines (Geyer and Daniilidis 2002a, 1999; Barreto and Araujo 2002, 2005; Ying and Hu 2004a; Ying and Zha 2005; Vandeportaele et al. 2006; Wu et al. 2006; Caglioti et al. 2007; Wu et al. 2008). Finally there are other methods that perform a self-calibration of the system (Kang 2000; Micusik and Pajdla 2006; Ramalingam et al. 2010; Espuny and Burgos Gil 2011).

As observed above there exist many calibration methods. They use different techniques and models to calibrate the omnidirectional systems. Some works have tried either to classify or to compare them. Ying and Hu (2004a) classify the calibration methods in three categories: (i) Known World Coordinates; (ii) Self-calibration: and (iii) Projection of lines. They only consider seven methods. On the other hand, they do not perform any comparison with any of such methods. A more specific classification is given by Deng et al. (2007) where 10 methods are grouped in five categories: (i) Self-calibration; (ii) Sphere-based calibration; (iii) Line-based calibration; (iv) Point-based calibration; and (v) 2D calibration. We observe that methods based on 2D patterns have appeared emulating calibration methods for conventional cameras. The sphere-based category only contains one method, which also uses lines and it could be classified in that category. In that work there is no comparison of the proposed method to any other. Frank et al. (2007) identify three big groups of calibration methods: (i) Known World Coordinates which include those based on 2D patterns, which from our point of view should belong to different categories; (ii) Geometric Invariants which include the methods based on projections of lines; and (iii) Self-calibration methods. They mention a total of eight methods. A comparison of the proposed method with the online available methods (Scaramuzza et al. 2006; Mei and Rives 2007) is presented. They use four different datasets including two fish-eye, a paracatadioptric system and a system with very small distortion. Since (Mei and Rives 2007) does not allow the manual extraction of grid points, the authors only consider those images where the grid points are extracted successfully, having a limited set of images. This situation has as consequence a poor calibration of the system. In this book we extract the grid points manually, allowing the methods to have the maximum data available, which permits to obtain the best performance and in consequence to perform a fair comparison of the methods. Toepfer and Ehlgen (2007) do not present a classification but a comparison of their proposed method with (Scaramuzza et al. 2006; Mei and Rives 2007; Tsai 1987). The performance of the methods is given considering a combination of the root-mean-square error (we assume the reprojection error) with the number of parameters of the method, through the principle of minimum description length. However, it is not clear which catadioptric system has been calibrated neither how the method (Tsai 1987) is adapted to work with catadioptric systems.

In this chapter, we first present a classification of the existing approaches to calibrate omnidirectional systems. We propose five categories based on the main entity required by each method to perform the calibration of the systems. We also present in Table 3.1 the relevant information of each method according to valuable criteria: pattern or entity required, considering the minimum number of elements, number of views, analytical model, type of mirror, or if the method requires the separate calibration of the mirror and the camera. A previous version of this comparison was presented in (Puig et al. 2012a).

Since the amount of calibration methods is considerable, the selection of a particular calibration method seems to be difficult and even more so if we consider the problems involved in the implementation process. Among all approaches mentioned previously, there is a group of calibration methods for omnidirectional systems (catadioptric and fish-eye) available online as OpenSource toolboxes. These methods can save time and effort when the goal is beyond the calibration itself and when the user is more interested in obtaining 3D motion and structure results than to deal with complex projection models. In this book, we evaluate these methods and provide an analysis with simulations and real images. Moreover, we use a structure from motion application with two omnidirectional images, where we become users of the calibrations provided by these approaches. This experiment shows the behavior of the approaches in a real scenario. Besides the performance, we also consider the ergonomics and ease of usage, as well as the type of features, the algorithm and the type of pattern, since they are important elements that can help the user to select the best approach. Moreover, we present an up-to-date list of the calibration methods developed that consider catadioptric and fish-eye systems, allowing the reader to decide to implement a different method. These calibration methods are:

1. Mei and Rives (2007)[1] which uses the sphere camera model and requires several images of a 2D pattern. We will call this approach *Sphere-2D Pattern*.
2. Puig et al. (2011)[2] which obtains a solution in closed form requiring a set of 3D to 2D correspondences. It also uses the sphere camera model. We call this approach *DLT-like*.
3. Barreto and Araujo (2002)[3] uses also the sphere camera model and requires a single omnidirectional image containing a minimum of three lines. We call it *Sphere-Lines*.
4. Scaramuzza et al. (2006)[4] which models the omnidirectional images as distorted images where the parameters of distortion have to be found. We call this approach *Distortion-2DPattern*.

---

[1] http://www.robots.ox.ac.uk/~cmei/Toolbox.html

[2] http://webdiis.unizar.es/~lpuig/DLTOmniCalibration

[3] http://www.isr.uc.pt/~jpbar/CatPack/main.htm

[4] http://asl.epfl.ch/~scaramuz/research/Davide_Scaramuzza_files/Research/OcamCalib_Tutorial.htm

Table 3.1 Classification of the calibration methods for omnidirectional systems

Method	Pattern/Entity	Number of views	Model	Mirror	W/S
Kang (2000)	10 point tracks	Multiple	Particular central catadioptric	Parabolic	W
Svoboda and Pajdla (2002)	–	Single	Particular central catadioptric	Generic	S
Caglioti et al. (2007)	1 line + mirror contours	Single	Particular noncentral catadioptric	Generic	W
Aliaga (2001)	3D known points	Multiple	Particular noncentral catadioptric	Parabolic	W
Toepfer and Ehlgen (2007)	3D pattern/2D pattern (multiple points)	Single/Multiple	Particular central catadioptric	Hyperbolic and parabolic	W
Geyer and Daniilidis (2002a)	3 lines	Single	Sphere	Parabolic	W
Geyer and Daniilidis (1999)	2 vanishing points	Single	Sphere	Parabolic	W
Barreto and Araujo (2005)	3 lines	Single	Sphere	Generic	W
Ying and Hu (2004a)	2 lines/3 spheres	Single	Sphere	Generic	W
Ying and Zha (2005)	3 lines	Single	Sphere	Generic	W
Vandeportaele et al. (2006)	3 lines	Single	Sphere	Parabolic	W
Wu et al. (2006)	lines	Single	Sphere	Parabolic	W
Wu et al. (2008)	3 lines	Single	Sphere	Generic	W
Vasseur and Mouaddib (2004)	lines (minimum no. n/a)	Single	Sphere	Generic	W
Mei and Rives (2007)	2D pattern (multiple points)	Multiple	Sphere	Generic	W
Puig et al. (2011)	3D pattern (20 3D–2D correspondences)	Single	Sphere	Generic	W
Deng et al. (2007)	2D pattern (multiple points)	Multiple	Sphere	Generic	W
Gasparini et al. (2009)	2D pattern (multiple points)	Multiple	Sphere	Generic	W
Wu and Hu (2005)	4 correspondences	Multiple	Sphere	Generic	W
Scaramuzza et al. (2006)	2D pattern (multiple points)	Multiple	Distortion	Generic	W
Frank et al. (2007)	2D pattern (multiple points)	Multiple	Distortion	Generic	W
Micusik and Pajdla (2006)	9 correspondences (epipolar geometry)	Multiple	Distortion	Generic	W
Morel and Fofi (2007)	3 polarized images	Multiple	Generic camera[a]	Generic	S
Ramalingam et al. (2010)	2 Rotational and Translational Flows	Multiple	Generic camera[a]	Generic	W
Espuny and Burgos Gil (2011)	2 Rotational Flows	Multiple	Generic camera[a]	Generic	W

W = whole system, S = separate calibration (1. camera and 2. mirror parameters)

[a] They use the same approach Sturm and Ramalingam (2004)

3.2 Classification of Calibration Methods for Omnidirectional Cameras

As observed above there exist many calibration methods. They use different techniques and models to calibrate the omnidirectional systems. Some of them first calibrate the perspective camera and after that find the mirror parameters. In this section, we present a review and classification of the existing calibration methods. We propose five categories based on the main entity required to perform the calibration of the systems.

Line-based calibration. Many methods are based on the projection of lines in the omnidirectional images. The main advantage of using lines is that they are present in many environments and a special pattern is not needed. These approaches compute the image of the absolute conic from which they compute the intrinsic parameters of the catadioptric system. Geyer and Daniilidis (2002a) calibrate para-catadioptric cameras from the images of only three lines. Ying and Hu (2004a) analyze the relation of the camera intrinsic parameters and imaged sphere contours. They use the projection of lines as well as projections of the sphere. The former gives three invariants and the latter two. Vasseur and Mouaddib (2004) detect lines in the 3D scene which are later used to calculate the intrinsic parameters. This approach is valid for any central catadioptric system. Vasseur and Mouaddib (2004) show that all line images from a catadioptric camera must belong to a family of conics which is called a line image family related to certain intrinsic parameters. They present a Hough transform for line image detection which ensures that all detected conics must belong to a line image family related to certain intrinsic parameters. Barreto and Araujo (2005) study the geometric properties of line images under the central catadioptric model. They give a calibration method suitable for any kind of central catadioptric system. Vandeportaele et al. (2006) slightly improve (Geyer and Daniilidis 2002a) using a geometric distance instead of an algebraic one and they allow to deal with lines that are projected to straight lines or to circular arcs in a unified manner. Wu et al. (2006) introduce a shift from the central catadioptric model to the pinhole model from which they establish linear constraints on the intrinsic parameters. Without conic fitting they are able to calibrate para-catadioptric-like cameras. Caglioti et al. (2007) calibrate a system where the perspective camera is placed in a generic position with respect to a mirror, i.e., a noncentral system. They use the image of one generic space line, from which they derive some constraints that, combined with the harmonic homology relating the apparent contours of the mirror allow them to calibrate the catadioptric system. More recently, Wu et al. (2008) derive the relation between the projection on the viewing sphere of a space point and its catadioptric image. From this relation, they establish linear constraints that are used to calibrate any central catadioptric camera.

2D pattern calibration. This kind of methods use a 2D calibration pattern with control points. These control points can be corners, dots, or any features that can be easily extracted from the images. Using iterative methods extrinsic and intrinsic parameters can be recovered. Scaramuzza et al. (2006) propose a technique to calibrate single viewpoint omnidirectional cameras. They assume that the image projection function can be described by a Taylor series expansion whose coefficients are estimated by solving a two-step least squares linear minimization problem. Mei and Rives (2007) propose as Scaramuzza a flexible approach to calibrate single viewpoint sensors from planar grids, but based on an exact theoretical projection function—the sphere model—to which some additional radial and tangential distortion parameters are added to consider real-world errors. Deng et al. (2007) use the bounding ellipse of the catadioptric image and the field of view (FOV) to obtain the intrinsic parameters. Then, they use the relation between the central catadioptric and the pinhole model to compute the extrinsic parameters. Gasparini et al. (2009) compute the image of the absolute conic (IAC) from at least three homographies which are computed from images of planar grids. The intrinsic parameters of the central catadioptric systems are recovered from the IAC.

3D Point-based calibration. These methods require the position of 3D points observed usually in a single image. Aliaga (2001) proposes an approach to estimate camera intrinsic and extrinsic parameters, where the mirror center must be manually determined. This approach only works for para-catadioptric systems. Wu and Hu (2005) introduced the invariants of 1D/2D/3D space points and use them to compute the camera principal point with a quasi-linear method. Puig et al. (2011) present an approach based on the Direct Linear Transformation (DLT) using lifted coordinates to calibrate any central catadioptric camera. It computes a generic projection matrix valid for any central catadioptric system. From this matrix, the intrinsic and extrinsic parameters are extracted in a closed form and refined by nonlinear optimization afterwards. This approach requires a single omnidirectional image containing points spread in at least three different planes.

Self-calibration. This kind of calibration techniques uses only point correspondences in multiple views, without needing to know either the 3D location of the points or the camera locations. Kang (2000) uses the consistency of pairwise tracked point features for calibration. The method is only suitable for para-catadioptric systems. Micusik and Pajdla (2006) propose a method valid for fish-eye lenses and catadioptric systems. They show that epipolar geometry of these systems can be estimated from a small number of correspondences. They propose to use a robust estimation approach to estimate the image projection model, the epipolar geometry, and to avoid outliers. Ramalingam et al. (2010) use pure translations and rotations and the image matches to calibrate central cameras from geometric constraints on the projection rays. Espuny and Burgos Gil (2011) developed a similar approach that uses two dense rotational flows produced by rotations of the camera about two unknown linearly independent axes which pass through the projection center.

Polarization Imaging. This method is proposed by Morel and Fofi (2007). It is based on an accurate reconstruction of the mirror by means of polarization imaging. It uses a very simple camera model which allows them to deal with any type of camera. However, they observe that developing an efficient and easy-to-use calibration method is not trivial.

3.3 Calibration Methods Analyzed

In this section, we summarize the four OpenSource methods used to calibrate omni-directional systems. The purpose of this section is to show a general view of the methods that help the reader to understand the core of each method.

3.3.1 Sphere-2D Pattern Approach

This approach (Mei and Rives 2007) uses the sphere model explained in the last section, with the difference that it does not consider the image flip induced by $(\psi - \xi)$, it uses $(\xi - \psi)$ in x and y coordinates. This approach adds to this model distortion parameters to consider real world errors. This method is multiview, which means that it requires several images of the same pattern containing as many points as possible. This method needs the user to provide prior information to initialize the principal point and the focal length of the catadioptric system. The principal point is computed from the mirror center and the mirror inner border. The focal length is computed from three or more collinear nonradial points. Once all the intrinsic and extrinsic parameters are initialized a nonlinear process is performed. This approach is also valid for fish-eye lenses and spherical mirrors.

This approach uses a total of 17 parameters to relate a scene point to its projection in the image plane:

- Seven extrinsic parameters $(\mathbf{q}, \mathbf{t})$ representing the relation between the camera reference system and the world reference system. A 4 vector $\mathbf{q}$ represents the rotation as a quaternion and a 3 vector $\mathbf{t}$ represents the translation.
- One mirror parameter ξ.
- Four distortion parameters $Dist = [k_1, k_2, p_1, p_2]$, two for tangential distortion and two for radial distortion (Heikkila and Silven 1997).
- Five parameters representing a generalized camera projection $\mathbf{P} = [\theta, \gamma_1, \gamma_2, u_0, v_0]$. (γ_1, γ_2) are the focal lengths of the catadioptric system for x and y axis, θ is the skew parameter, and (u_0, v_0) is the principal point.

The 2D pattern used to calibrate the camera is composed of m points $\mathbf{X}_i$ with their associated image values $\mathbf{x}_i$. The solution of the calibration problem is obtained by minimizing the reprojection error using the Levenberg–Marquardt algorithm.

3.3.2 Sphere-Lines Approach

This method (Barreto and Araujo 2005) is based on computing the absolute conic $\hat{\Omega}_\infty = H_c^{-T} H_c^{-1}$ and the mirror parameter ξ under the sphere camera model. In omnidirectional images 3D lines are mapped to conics. So the first step is to fit these conics. With the information provided by these conics and the location of the principal point an intermediate step is performed. It computes entities like polar lines, lines at infinity, and circle points. From these intermediate entities and some invariant properties like collinearity, incidence, and cross-ratio the mirror parameter ξ is computed. From the image of a conic in the omnidirectional images, it is possible to compute two points that lie on the image of the absolute conic. Since a conic is defined by a minimum of five points at least three conic images are required to obtain $\hat{\Omega}_\infty$. Once the image of the absolute conic $\hat{\Omega}_\infty$ is computed, from its Cholesky decomposition we obtain

$$H_c = \begin{pmatrix} \gamma_x & \beta & c_x \\ 0 & \gamma_y & c_y \\ 0 & 0 & 1 \end{pmatrix} \tag{3.1}$$

with the intrinsic parameters γ_x and γ_y (focal lengths), skew (β) and principal point (c_x, c_y).

3.3.3 DLT-Like Approach

This approach (Puig et al. 2011) also uses the sphere camera model. To deal with the nonlinearities present in this model, the lifting of vectors and matrices is used. This method computes a lifted 6×10 projection matrix that is valid for all single-viewpoint catadioptric cameras. The required input for this method is a single image with a minimum of 20 3D–2D correspondences distributed in three different planes.

In this approach, a 3D point $\mathbf{X}$ is mathematically projected to two image points $\mathbf{x}_+$, $\mathbf{x}_-$, which are represented in a single entity via a degenerate dual conic Ω. The relation between them is $\Omega \sim \mathbf{x}_+\mathbf{x}_-^\mathsf{T} + \mathbf{x}_-\mathbf{x}_+^\mathsf{T}$.

This conic represented as a 6 vector $\mathbf{c} = (c_1, c_2, c_3, c_4, c_5, c_6)^\mathsf{T}$ projected on the omnidirectional image is computed using the lifted 3D-point coordinates, intrinsic and extrinsic parameters as:

$$\mathbf{c} \sim \widehat{H}_{c6\times 6} X_\xi \widehat{R}_{6\times 6} \left(I_6 \ T_{6\times 4} \right) \widehat{\mathbf{X}}_{10} \tag{3.2}$$

where, R represents the rotation of the catadioptric camera. X_ξ a 6×6 matrix and $T_{6\times 4}$ depend only on the sphere model parameter ξ and position of the catadioptric camera $\mathbf{C} = (t_x, t_y, t_z)$ respectively. Thus, a 6×10 catadioptric projection matrix, P_{cata}, is expressed by its intrinsic A_{cata} and extrinsic T_{cata} parameters

$$P_{\text{cata}} = \underbrace{\widehat{H}_c X_\xi}_{A_{\text{cata}}} \underbrace{\widehat{R}_{6\times 6} \left(I_6 \ T_{6\times 4} \right)}_{T_{\text{cata}}} \tag{3.3}$$

This matrix is computed from a minimum of 20 3D–2D lifted correspondences in a similar way to the perspective case (Hartley and Zisserman 2000) using least squares

$$\left([\widehat{\mathbf{x}}]_\times \otimes \widehat{\mathbf{X}}\right) \mathbf{p}_{\text{cata}} = \mathbf{0}_6 \tag{3.4}$$

The 60 vector $\mathbf{p}_{\text{cata}}$ contains the 60 coefficients of P_{cata}. Manipulating this matrix algebraically the intrinsic and extrinsic parameters are extracted. These extracted values are used as an initialization to perform a nonlinear process (Levenberg–Marquardt). In this process, some parameters that are not included in the sphere camera model are added. These parameters are, as in Sphere-2D Pattern approach, the radial and tangential distortion which are initialized to zero. This approach uses two parameters for each type of distortion. The minimization criterion is the root-mean square (RMS) of distance error between a measured image point and its reprojected correspondence.

3.3.4 Distortion-2D Pattern Approach

In this approach, (Scaramuzza et al. 2006) the projection model is different from the one previously presented. The only assumption is that the image projection function can be described by a polynomial, based on Taylor series expansion, whose coefficients are estimated by solving a two-step least squares linear minimization problem. It does not require either any a priori knowledge of the motion or a specific model of the omnidirectional sensor. Hence, this approach assumes that the omnidirectional image is in general a highly distorted image and we have to compute the distortion parameters to obtain such a distorted image. This approach, like Sphere-2D Pattern, requires several images from different unknown positions of a 2D Pattern. The accuracy depends on the number of images used and on the degree of the polynomial.

Under this model, a point in the camera plane $\mathbf{x}' = [x', y']^\mathsf{T}$, is related to a vector $\mathbf{p}$ which represents a ray emanating from the viewpoint O (located at the focus of the mirror) to the scene point $\mathbf{X}$. This relation is encoded in the function $\mathbf{g}$

$$\mathbf{p} = \mathbf{g}(x') = \mathsf{P}\mathbf{X} \tag{3.5}$$

where $\mathbf{X} \in \Re^4$ is expressed in homogeneous coordinates; $\mathsf{P} \in \Re^{3 \times 4}$ is a perspective projection matrix. The function $\mathbf{g}$ has the following form

$$\mathbf{g}(\mathbf{x}') = (x', y', f(x', y'))^\mathsf{T}, \tag{3.6}$$

and f is defined as

$$f(x', y') = a_0 + a_1 \rho' + a_2 \rho'^2 + \cdots + a_n \rho'^n \tag{3.7}$$

where ρ' is the Euclidean distance between the image center and the point. In order to calibrate the omnidirectional camera the $n + 1$ parameters $(a_0, a_1, a_2, \ldots, a_n)$ corresponding to the coefficients of function f need to be estimated.

The camera calibration under this approach is performed in two steps. The first step computes the extrinsic parameters, i.e., the relation between each location of the planar pattern and the sensor coordinate system. Each point on the pattern gives three homogeneous equations. Only one of them is linear and it is used to compute the extrinsic parameters. In the second step, the intrinsic parameters are estimated, using the extrinsic parameters previously computed and the other two equations. The authors do not mention it, but after this linear process a nonlinear optimization is performed using the Gauss–Newton algorithm.[5]

3.4 Experiments

In order to compare the different calibration methods we calibrate a hypercatadioptric system,[6] a fish-eye, and a commercial proprietary shape mirror,[7] which we name unknown-shape system. Additionally, we displace the perspective camera of the hypercatadioptric system far from the mirror. This allows the displacement of the optical center of the perspective camera from the other focus described by the hyperbolic mirror, leading to a noncentral system. We calibrate these four systems with the four methods and compare the results with a reconstruction experiment which is explained in Sect. 3.4.1. The setup used to calibrate the omnidirectional system for every method is explained as follows.

Sphere-2D Pattern approach. This approach requires images of a single 2D pattern. These images have to cover most of the omnidirectional image area. This approach asks the user for the image center and for the outer mirror border to compute the principal point. Then it asks for four aligned edge points on a nonradial line to compute the focal length. With this information it asks for the four corner points of the pattern and uses a subpixel technique to extract the rest of the points present in the pattern. If the focal length is not well estimated then all points have to be given manually.

DLT-like approach. In the DLT-like approach a single image of a 3D pattern was used. This approach does not have an automatic extractor so all points are given manually. This method requires as input a set of points lying on at least three different planes with known relative position.

Distortion-2D approach. This approach also uses images coming from a 2D pattern. The last version of this method has an automatic corner detector which detects most of the corners present in a single pattern. The amount of corners given manually

---

[5] This algorithm is provided by the **lsqnonlin** Matlab function.

[6] Neovision H3S with XCD-X710 SONY camera.

[7] http://www.0-360.com

is minimum. Once all the points in all the images are given the calibration process starts.

Sphere-Lines approach. This approach is based on the projections of lines in the omnidirectional images. This method only requires one omnidirectional image containing at least three lines.

3.4.1 Evaluation Criteria

Previous to the comparison of the real system we perform an experiment using simulated data. The purpose of this experiment is to observe the behavior of all approaches under optimal conditions and to measure their sensitivity to noise. We simulate two central catadioptric systems, a hypercatadioptric[8] with mirror parameter $\xi = 0.7054$ and a paracatadioptric system. First, we generate five synthetic images, each one containing a calibration pattern, that covers the whole FOV of the omnidirectional image, two of these images are shown in Fig. 3.1a, b. These images are used by the Sphere-2D Pattern and the Distortion-2D approaches. We combine the five calibration patterns in a single image (Fig. 3.1c) that is used by the DLT-Like approach. These three approaches use the same points. In the case of the Sphere-Lines approach four lines are present in a single image with a total of 904 points (Fig. 3.1d). We add Gaussian noise to the image coordinates. For every noise level σ (in pixels) we repeat the experiment 100 times in order to avoid particular cases due to random noise. In Fig. 3.2 and Fig. 3.3 we show the mean and the standard deviation of the reprojection error corresponding to the hypercatadioptric and the paracatadioptric systems, respectively for the analyzed approaches.

As we observe, the behavior of the four approaches is quite similar. All of them respond in the same way to the amount of noise present in the images. This experiment shows that under optimal circumstances the performance of the four approaches is quite similar. One thing we observe with respect to the Sphere-Lines approach is that four lines are enough to compute the calibration. We try to calibrate the systems using a higher number of lines which caused the approach being slower and in some occasions it did not converge. This behavior is due to the lines containing a high number of points; therefore increasing two or more lines means to increase hundreds of points.

We observe that the reprojection error is not sufficient to decide which method is endowed with the best performance; however, it is a necessary condition to qualify an algorithm as performing well. Moreover, a disadvantage of this error measure is that we can make it smaller by just adding more parameters to a model. As an example, we can observe the reprojection error of the hypercatadioptric system given by the Sphere-2D Pattern approach (see Table 3.3). The obtained reprojection error during calibration was considerably small, only 0.000005 pixels, which could be considered as zero. This approach adds five distortion parameters: three for radial

---

[8] http://www.accowle.com

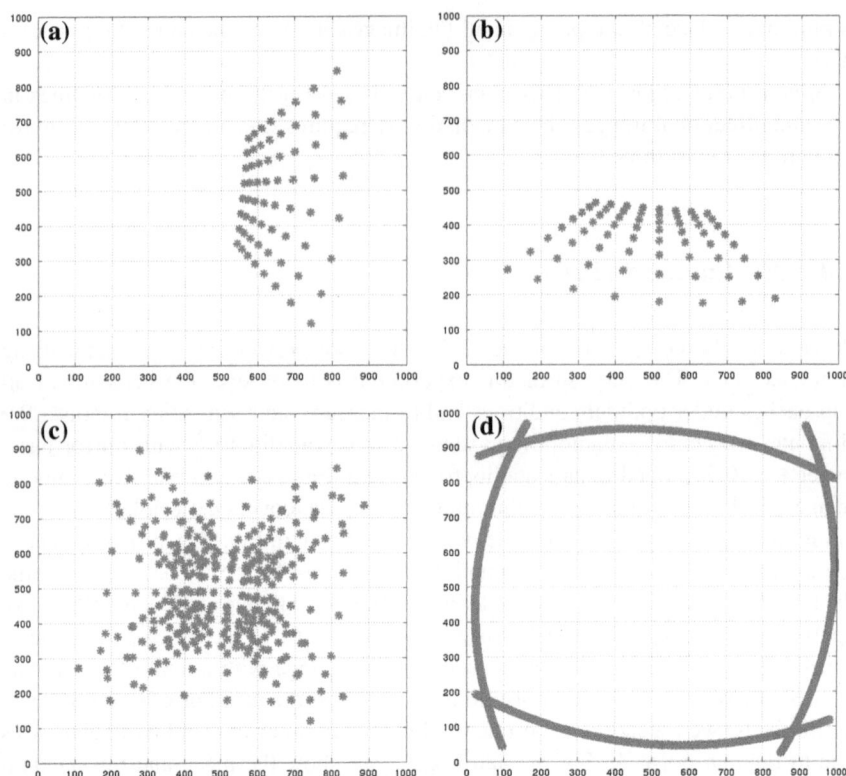

Fig. 3.1 Synthetical images used by the calibration methods. Sphere-2D Pattern and Distortion-2D Pattern approaches use five images similar to (**a**) and (**b**). DLT-like approach uses a single image containing five planar patterns (**c**) and Sphere-Lines approach use four image lines (**d**)

distortion and two for tangential distortion. Originally, these parameters are not considered in the model. To verify the impact of these parameters in the calibration we repeated the calibration. When we only consider radial distortion (3 parameters), the reprojection error increased to 0.42 pixels. When no distortion is considered at all, the reprojection error increased to 0.64 pixels. As previously stated, the reprojection error is not definitive in showing which approach performs the best.

Instead of the reprojection error, we choose to consider a structure from motion task from two calibrated omnidirectional images of a 3D pattern (Fig. 3.4) built in our lab. The pattern has been measured with high accuracy using photogrammetric software.[9] Thus, a 3D reconstruction by bundle adjustment has been made. We used six convergent views taken with a calibrated high-resolution camera (Canon EOS 5D with 12.8 Megapixel). The estimated accuracy of the location of the 3D points is better than 0.1 mm. Figure 3.5 shows the configuration used for the SfM experiment. Using the calibration provided by each method we compute the corresponding 3D

---

[9] PhotoModeler software was used.

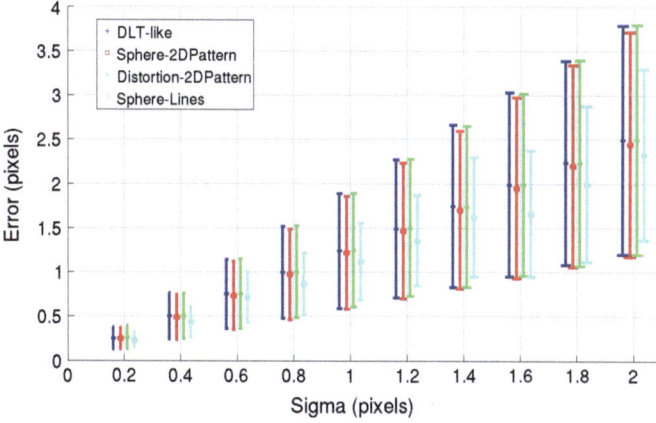

Fig. 3.2 Reprojection error in pixels as a function of noise (σ) corresponding to the hypercatadioptric system

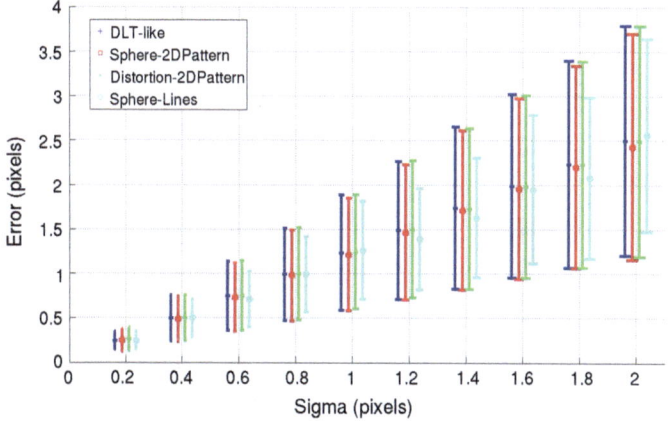

Fig. 3.3 Reprojection error in pixels as a function of noise (σ) corresponding to the paracatadioptric system

rays from each omnidirectional image. The correspondences between the two images were given manually. We use these correspondences to compute the essential matrix E which relates them. From this matrix we compute two projection matrices $\mathsf{P}_1 = [\mathsf{I}|\mathbf{0}]$ and $\mathsf{P}_2 = [\mathsf{R}|\mathbf{t}]$. Then, with these projection matrices and the 3D rays we compute an initial 3D reconstruction using a linear triangulation method (Hartley and Zisserman, 2000) which is later refined by a bundle adjustment optimization process. The 3D reconstruction depends on the number of correspondences. We use a set of 144 points to compute the reprojection error and to evaluate the 3D reconstruction results. We choose two different criteria to measure the accuracy of each model. These criteria are:

Fig. 3.4 3D pattern built at lab

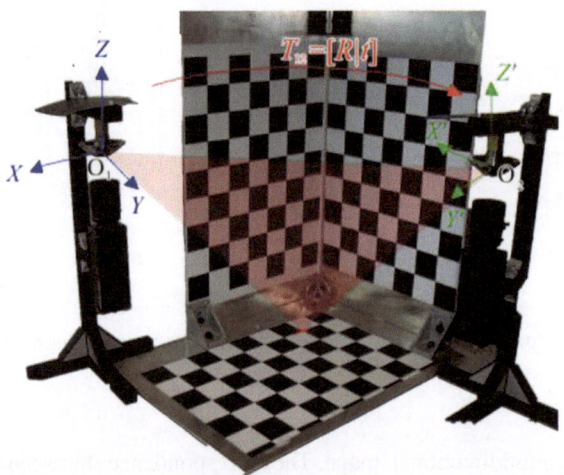

Fig. 3.5 SfM configuration to test the calibrations

- The average error between the real 3D points and their estimations.
- The reprojection error. We project the ground truth 3D pattern in the two cameras with the locations given by the SfM algorithm. We measure the error in pixels between the image points and the ground truth reprojection.

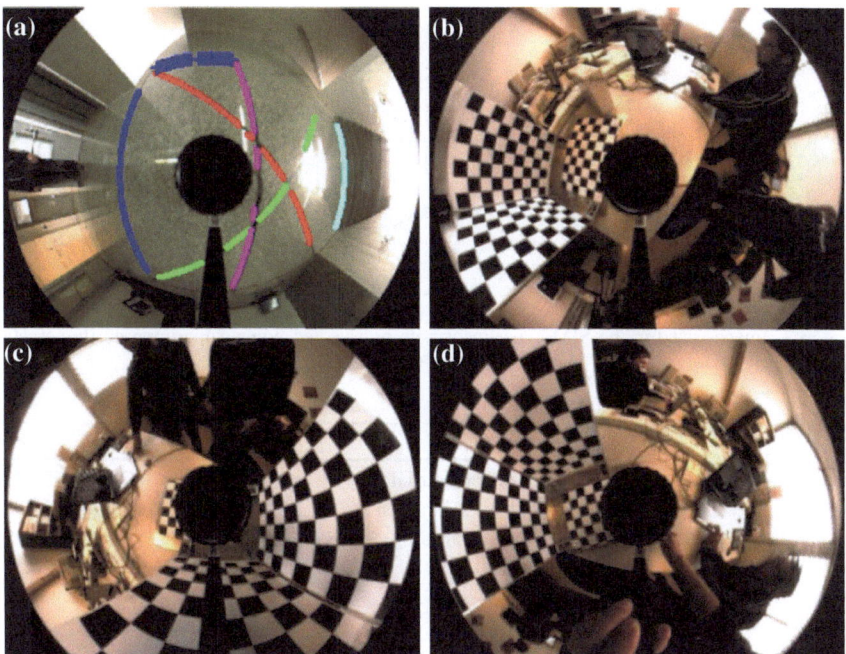

Fig. 3.6 Some images used to calibrate the hypercatadioptric system. **a** Sphere-Lines. **b** DLT-like approach. **c,d** Sphere-2D Pattern and Distortion-2D Pattern approaches

3.4.2 Hypercatadioptric System

The hypercatadioptric system is composed by a perspective camera with a resolution of 1024×768 and a hyperbolic mirror having a 60 mm diameter and parameters $a = 281$ mm and $b = 234$ mm according to manufacturer information. The mirror parameter for the sphere camera model is $\xi = 0.9662$. In Fig. 3.6 we observe some of the images used to calibrate this system. We use one image to calibrate the system using the Sphere-Lines and the DLT-like approaches. Eleven images of the 2D pattern were used to calibrate the system using both Sphere-2D Pattern and Distortion-2D approaches. In Fig. 3.7 we show the images used for the SfM experiment.

3.4.2.1 Mirror Parameter and Principal Point

Three of these methods are based on the sphere camera model. In Table 3.2 we present the estimations of the principal point (u_0, v_0) and the mirror parameter ξ since they are related with sphere model parameters. The distortion-2D pattern approach does not offer information about the catadioptric system. As we can see, the best estimation of the mirror parameter is given by Sphere-2D Pattern but also the DLT-like

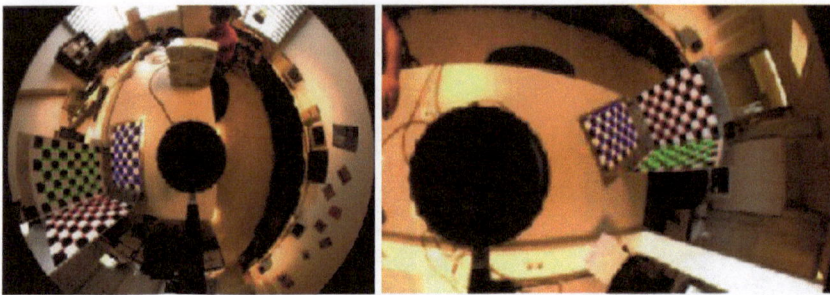

Fig. 3.7 Images used in the SfM experiment with reprojected points superimposed (hypercatadioptric)

Table 3.2 Comparison of the physical parameters given by the three methods based on the sphere model

	ξ	(u_0, v_0)
Manufacturer info	0.9662	(511.88, 399.25)
Sphere-2D Pattern	0.9684	(513.93, 400.76)
Sphere-Lines	1.0215	(523.82, 416.29)
DLT-like	0.9868	(509.95, 398.54)

algorithm gives a good approximation. Sphere-Lines give a value bigger than one which does not correspond to a hyperbolic mirror. With respect to the principal point the estimation of Sphere-2D Pattern and DLT-like approaches are close to the real one. The difference is that the Sphere-2D Pattern method asks the user to give the image center and the rim of the mirror to compute the principal point and the DLT-like algorithm does not need any of this a priori information but requires known positions of the three planes in the pattern.

In Fig. 3.8 we show the histogram corresponding to the accuracy in millimeters of the 144 reconstructed 3D points. We observe that the Sphere-2D Pattern approach has the highest number of reconstructed points closer to the ground truth. The highest error corresponds to Sphere-Lines and Distortion-2D Pattern with one point 5 mm far from the ground truth. In Fig. 3.9 we show the reprojection error of the 288 points of the two images used in the experiment. We observe that all the approaches are below the 2 pixel error and three of them within the 1 pixel error.

The hypercatadioptric system is the more complex catadioptric system under the sphere camera model since the mirror parameter ξ is in the interval [0, 1]. In opposition, the paracatadioptric system where $\xi = 1$ simplifies considerably the model. In this context, we decide to present the reprojection error corresponding to this system from three different sources (see Table 3.3): the reprojection error shown in the original paper where the approach was firstly presented (first column), the reprojection error obtained when we calibrate the system, and the reprojection error from the structure from motion experiment. This information provides the reader with a clear idea about the performance of all approaches under different circumstances and allows us to observe that the reprojection error given at the calibration time is

Fig. 3.8 Number of reconstructed 3D points within an error in mm using a hypercatadioptric system

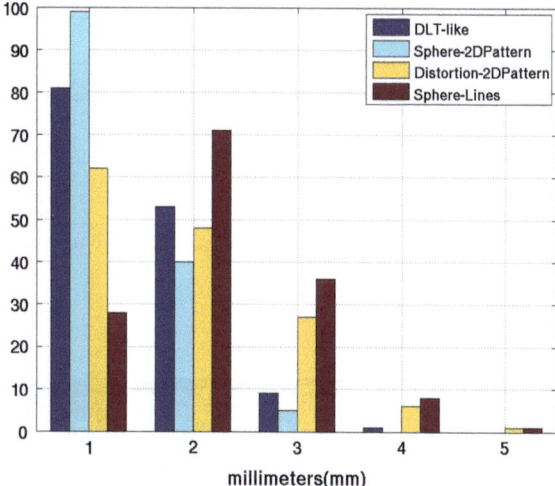

Fig. 3.9 Number of reprojected points within an error distance in pixels using a hypercatadioptric system

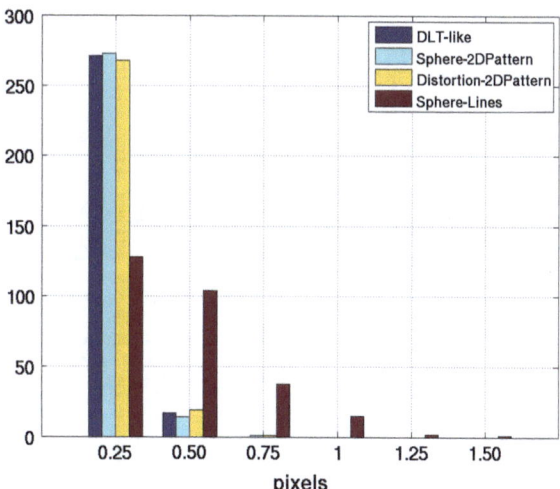

less informative that the reprojection error of a structure from motion experiment where the calibration is an early step and all the approaches are under the same circumstances.

3.4.3 Fish-Eye Lens

The fish-eye lens used in this experiment is a Raynox DCR-CF185PRO with a FOV of 185° on all directions. It is mounted on a high-resolution camera. In Fig. 3.10 we show

Table 3.3 Reprojection error from different sources corresponding to the hypercatadioptric system

Method	Original paper	Calibration	Structure from motion
Sphere-2D Pattern	0.40	0.00005	0.34
Sphere-Lines	n/a	1.11*	1.19
DLT-Like	0.30	0.47	0.40
Distortion-2D Pattern	1.2	0.82	0.45

(Since the Sphere-Lines does not use the reprojection error we take it from simulations with Gaussian noise $\sigma = 1$ pixel)

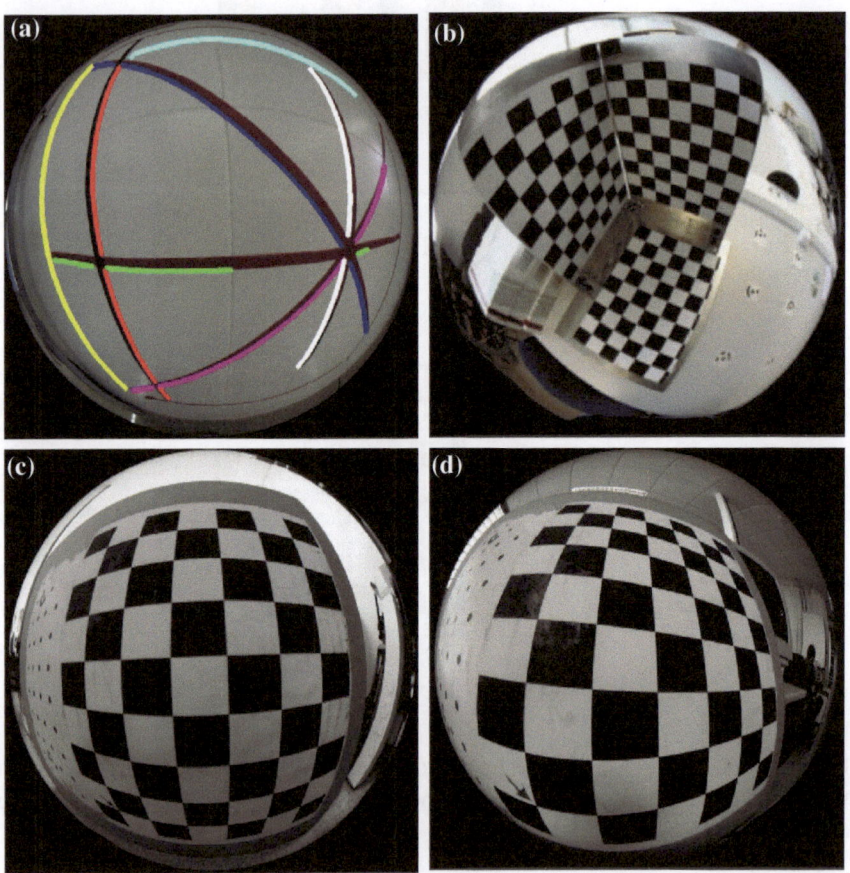

Fig. 3.10 Some images used to calibrate the fish-eye system. **a** Sphere-Lines. **b** DLT-like approach. **c,d** Sphere-2D Pattern and Distortion-2D Pattern approaches

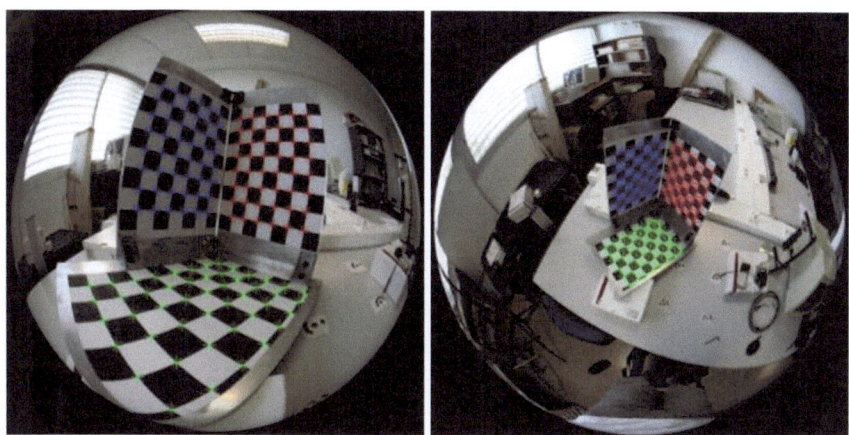

Fig. 3.11 Images used in the SfM experiment with reprojected points superimposed (fish-eye)

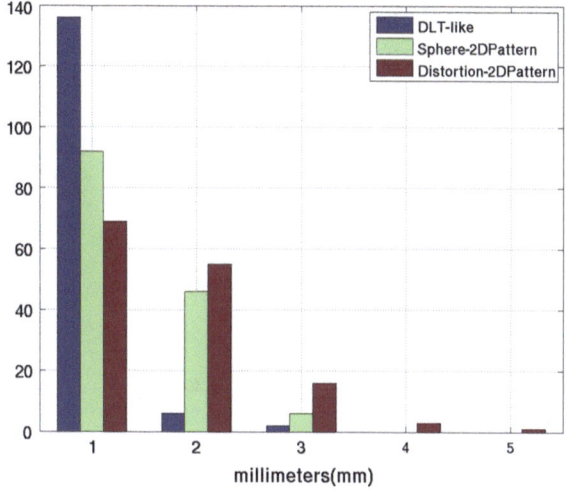

Fig. 3.12 Number of reconstructed 3D points within an error in mm using a fish-eye lens

some of the images used to calibrate this system. We use eight images to perform the calibration with the Sphere-2D Pattern and Distortion-2D Pattern approaches and only one image using the DLT-like approach. We use the image of seven lines to perform the calibration with the Sphere-Lines approach. Since none of the methods provides any information about the system we just show the results obtained from the SfM experiment. Figure 3.11 shows the images used for this experiment.

In Fig. 3.12 and Fig. 3.13 we show the results of this experiment. We observe that the methods that claim to be able to calibrate the slightly noncentral fish-eye systems indeed give good results. This is the case of the Sphere-2D Pattern, DLT-like and the Distortion-2D Pattern approaches. The opposite case is the Sphere-Lines approach

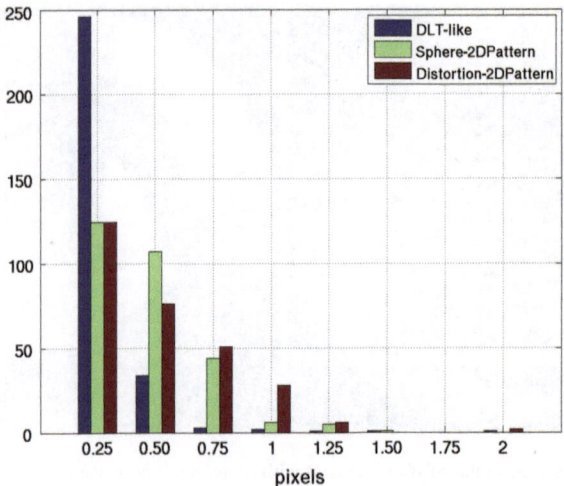

Fig. 3.13 Number of reprojected points within an error distance in pixels using a fish-eye lens

which is not able to correctly calibrate this camera. It gives reprojection errors close to 100 pixels. We observe that the three valid approaches give similar results. The DLT-like gives the best results with the highest number of reconstructed points within 1 mm error, although it is not designed to handle fish-eye lenses. With respect to the reprojection error, we observe a similar behavior of the three approaches with a maximum error of 2 pixels.

3.4.4 Unknown-Shape Catadioptric System

This system is the combination of a commercial proprietary shape mirror and a high-resolution camera. We use six images of a 2D pattern to calibrate this system for the Sphere-2D Pattern and Distortion-2D Pattern approaches. We use an image with four lines to perform the calibration with the Sphere-Lines approach. We observed several difficulties when more than four lines were used to calibrate the system using the Sphere-Lines approach. Sometimes the calibration results contained complex numbers or there were problems of convergence. In Fig. 3.14 we observe some of the images used to calibrate this system under all the analyzed approaches. In Fig. 3.15 we show the images used to perform the SfM experiment.

We decided to use a more complex reconstruction scenario to observe the behavior of the approaches under these conditions (Fig. 3.15a, b). The results of the experiments, the 3D accuracy and the reprojection error are shown in Fig. 3.16 and Fig. 3.17, respectively. We observe that the Distortion-2D Pattern approach has the highest number of points within the 5 mm error, 131 out of 144. The worst performance is given by the Sphere-Lines approach with maximum errors of 35 mm. The other

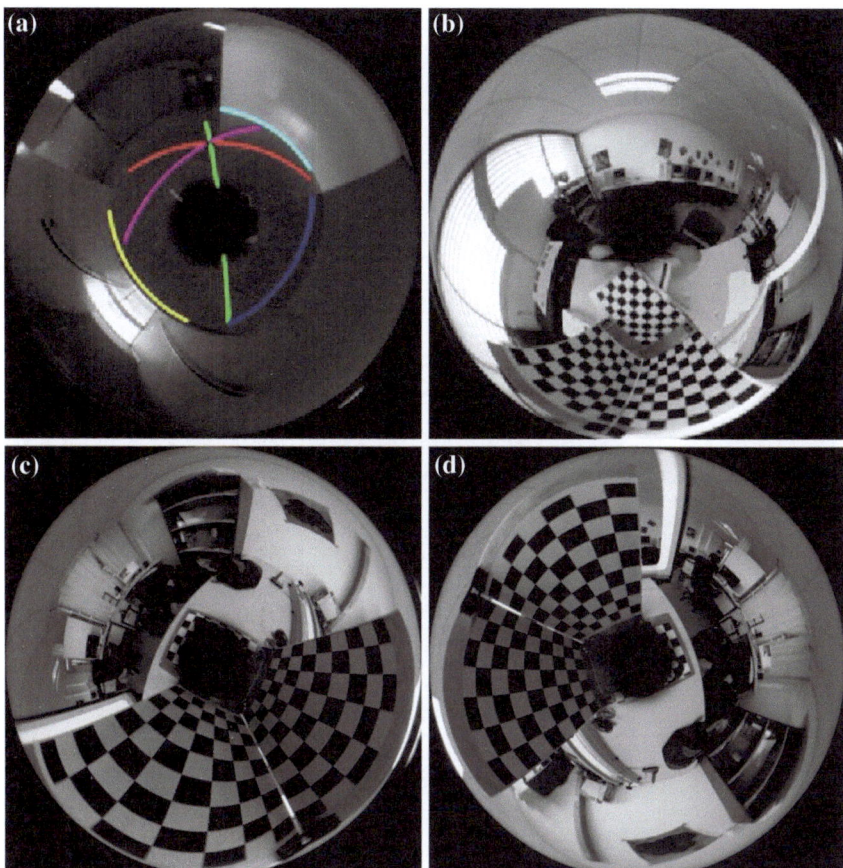

Fig. 3.14 Some images used to calibrate the unknown-shape catadioptric system. **a** Sphere-Lines. **b** DLT-like approach. **c,d** Sphere-2D Pattern and Distortion-2D Pattern approaches

two approaches show similar performance, with the majority of the reconstructed 3D points within the 10 mm error. The lowest reprojection error is given by the Distortion-2D Pattern with all the points within the 1 pixel error and the other three approaches have a similar behavior with the majority of the reprojected points within the 3 pixel error.

3.4.5 NonCentral Hypercatadioptric System

This system is the one described in the hypercatadioptric experiment. The only difference is that the perspective camera is displaced as far as possible from the mirror. This causes that the optical center of the perspective camera is not located

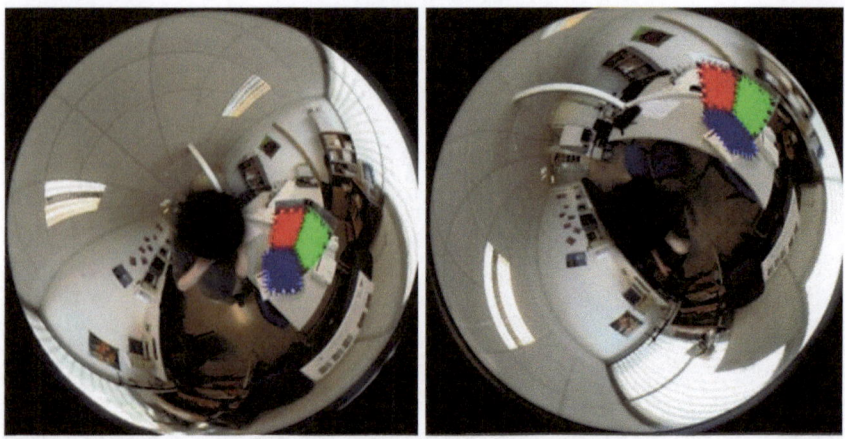

Fig. 3.15 Images used in the SfM experiment with reprojected points superimposed (unknown-shape)

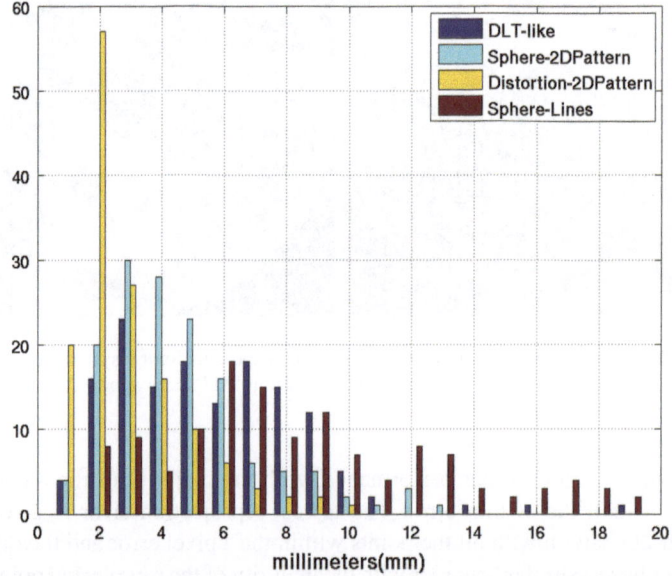

Fig. 3.16 Number of reconstructed 3D points within an error in mm using an unknown-shape system

at the other focus of the hyperbolic mirror, which is the basic condition for this system to be central. Some of the images used to perform the calibration of this system under the different models are shown in Fig. 3.18. As in the hypercatadioptric case we compute the mirror parameter and the principal point. This result is shown

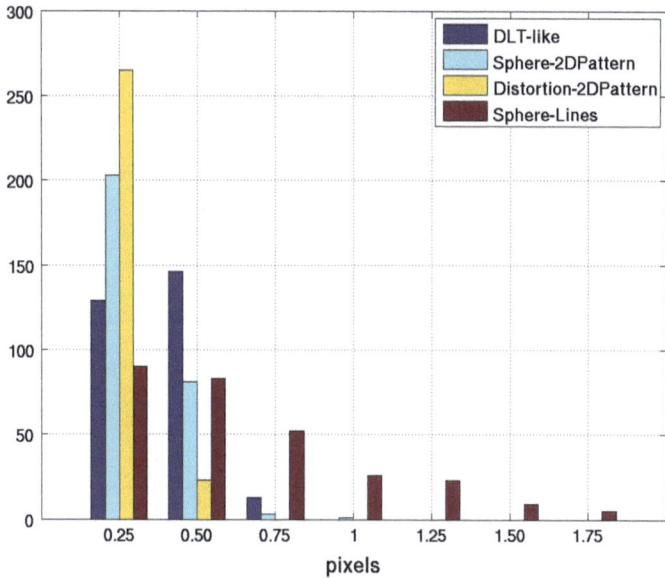

Fig. 3.17 Number of reprojected points within an error distance in pixels using an unknown-shape system

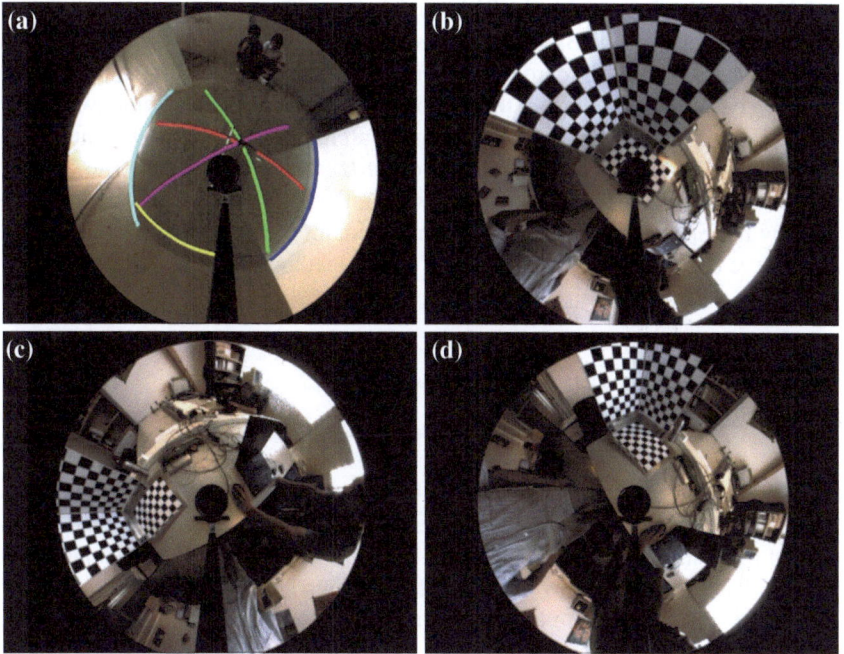

Fig. 3.18 Some images used to calibrate the noncentral catadioptric system. **a** Sphere-Lines. **b** DLT-like approach. **c,d** Sphere-2D Pattern and Distortion-2D Pattern approaches

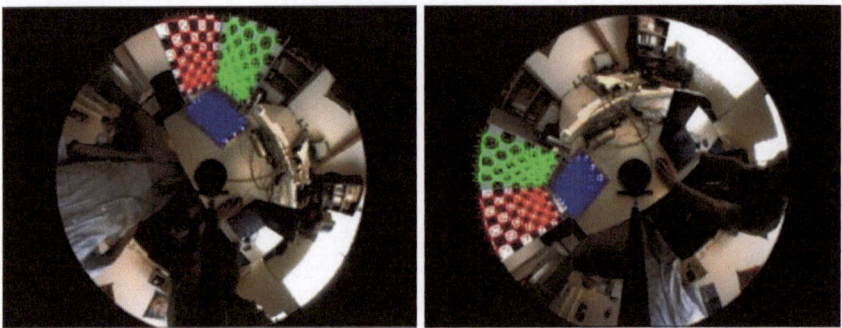

Fig. 3.19 Images used in the SfM experiment with reprojected points superimposed (noncentral hypercatadioptric)

Table 3.4 Comparison of the physical parameters given by the three methods based on the sphere model in the noncentral system

	ξ	(u_0, v_0)
Ground truth	0.9662	(511.88, 399.25)
Sphere-2DPattern	0.8463	(519.14, 407.87)
Sphere-Lines	1.00	(537.50, 409.82)
DLT-like	0.8819	(525.05, 411.89)

in Table 3.4. We observe that DLT-like and Sphere-2D Pattern give similar mirror hyperbolic parameter and Sphere-Lines estimates a parabolic mirror with $\xi = 1$.

The 3D error and the reprojection error are shown in Fig. 3.20 and Fig. 3.21, respectively. We observe that all the approaches have a similar behavior even with a noncentral system. The Sphere-2D Pattern has all its reconstructed 3D points within the three pixel error. The DLT-like and Distortion-2D Pattern approaches show similar results with one 3D reconstruction error within the 5 mm error. The worst result is given by the Sphere-Lines approach with maximum reconstruction error of 8 mm. We observe that the reprojection error for all the methods is below 2 pixels.

3.4.6 Distribution of the Calibration Points in the Catadioptric Image

The accuracy of the computed calibration relies on the area occupied by the calibration points in the calibration images. In this experiment, we show the importance of the distribution of the calibration points inside the catadioptric images. We define the area to be used by selecting those points closer than r pixels from the image center. The system to calibrate is a central hypercatadioptric system. Since the images used by each approach are not the same, and also the approaches use different features (points and lines), a full comparison using the same distances for all the approaches is not possible. Since the approaches Distortion-2D Pattern and

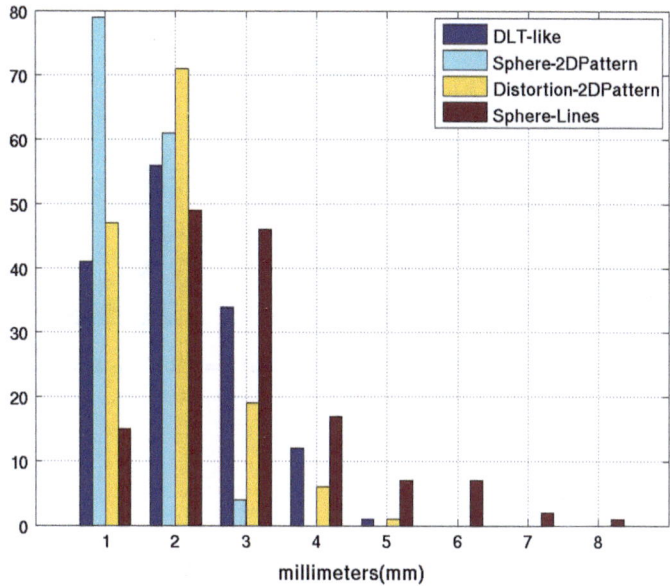

Fig. 3.20 Number of reconstructed 3D points within an error distance in millimeters using a noncentral hypercatadioptric system

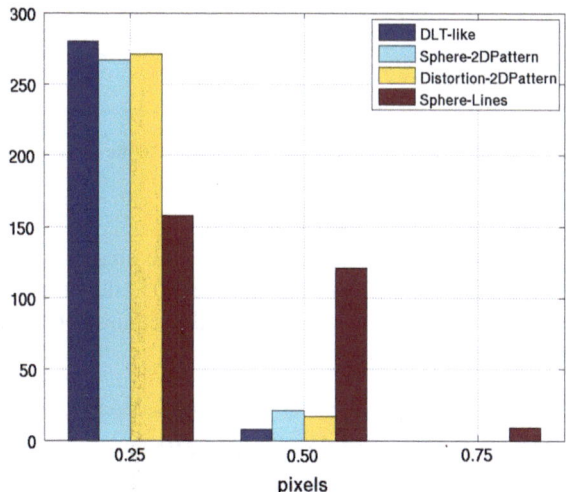

Fig. 3.21 Number of reprojected image points within an error distance in pixels using a noncentral hypercatadioptric system

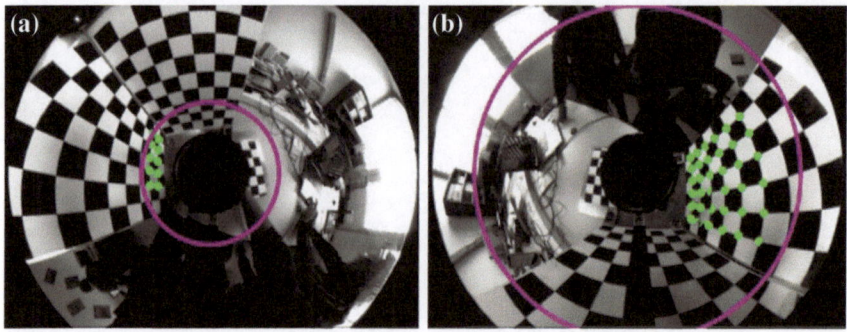

Fig. 3.22 Points contained inside the areas described by (**a**) $r = 150$ pixels and (**b**) $r = 400$ pixels

Fig. 3.23 3D reconstruction error in mm. The horizontal axis represents the different radii r

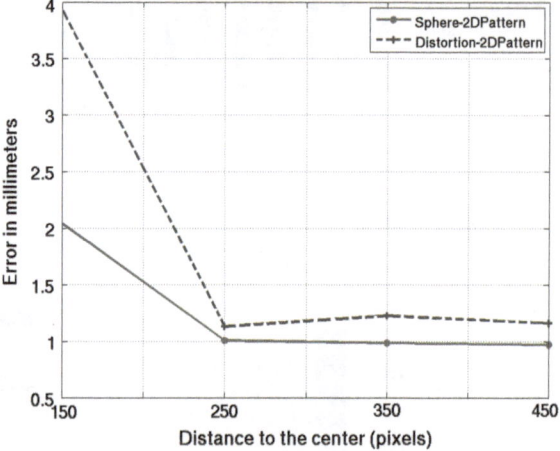

Sphere-2D Pattern share the calibration images set we present their corresponding results together. In the case of DLT-like and Sphere-2D Pattern the results are shown separately. The radii r were chosen depending on the requirements of each approach to calibrate the catadioptric system.

3.4.6.1 Distortion-2D Pattern and Sphere-2D Pattern

These two approaches require several images to perform the calibration. We select the calibration points that lie closer than r pixels from the image center in all the images of the set in which these points exist. An example for two different radii r can be observed in Fig. 3.22. In Fig. 3.23 we show the mean of the reconstruction error for each calibration performed with the points within the areas described by the radii r. We observe that both approaches give similar results. When the area is small, the number of points decreases and the estimation is worse. The Distortion-2D Pattern

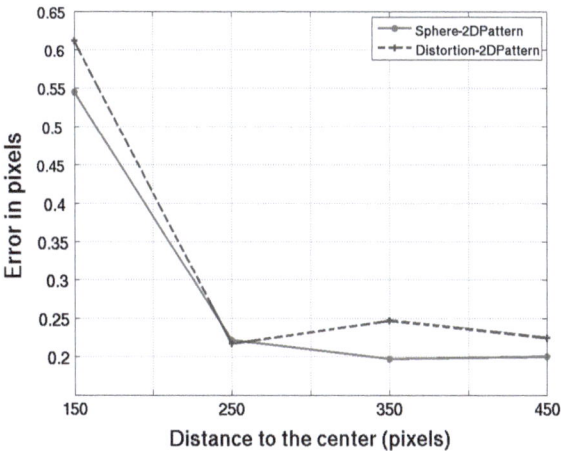

Fig. 3.24 Reprojection error in pixels. The horizontal axis represents the different radii r

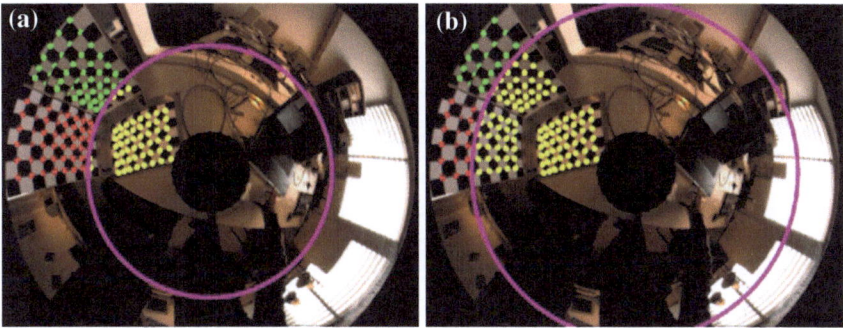

Fig. 3.25 Points contained inside the areas described by (**a**) $r = 300$ pixels and (**b**) $r = 400$ pixels

has an error of 4 mm and the Sphere-2D Pattern 2 mm in the worst case. This behavior can be explained by the fact that Sphere-2D Pattern estimates the image center from data given by the user and the Distortion-2D Pattern does it automatically, depending more on the distribution of the points. In Fig. 3.24 we show that reprojection error that is under 1 pixel error for both approaches, even with the smallest r.

3.4.6.2 DLT-Like

Since the DLT-like approach only requires one single image, the points are selected directly using different radii r. In Fig. 3.25 we show two examples of the points selected for two different radii. In Fig. 3.26 and Fig. 3.27 we show the 3D reconstruction error and the reprojection error, respectively. We observe a similar behavior to

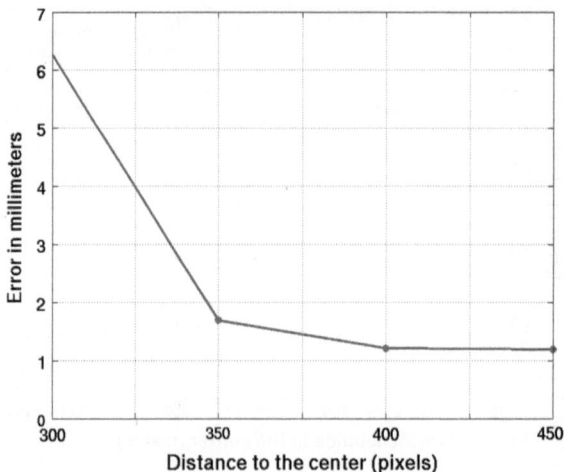

Fig. 3.26 3D reconstruction error in mm. The horizontal axis represents the different radii r

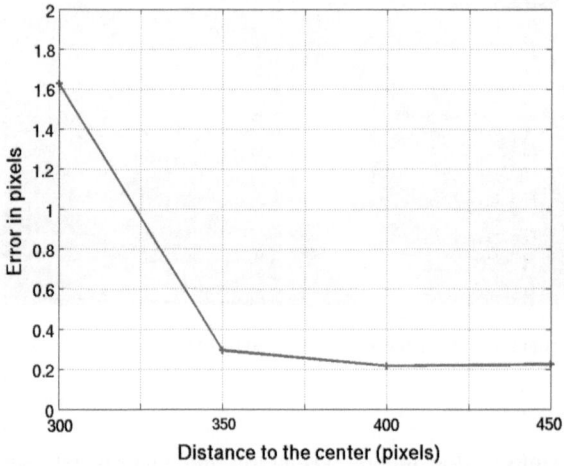

Fig. 3.27 Reprojection error in pixels. The horizontal axis represents the different radii r

the previous approaches using large radii. With small radii the results are worse, since with small radii only very few points of the second and third planes are considered (see Fig. 3.25a).

3.4.6.3 Sphere-Lines

This approach also requires a single image. The image must contain at least three line images. The radii limit the length of the lines used to calibrate the system. An

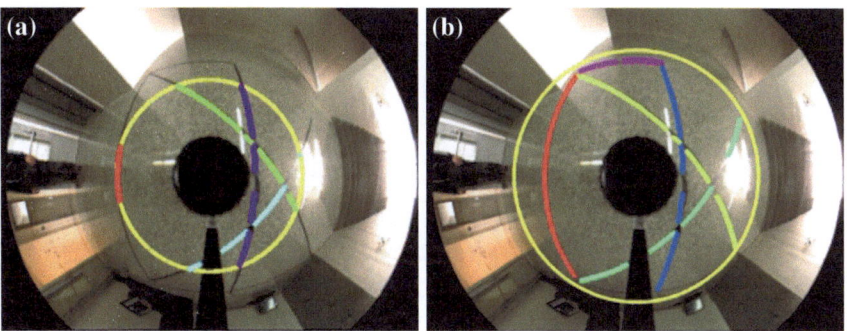

Fig. 3.28 Points contained inside the areas described by (**a**) $r = 230$ pixels and (**b**) $r = 300$ pixels.

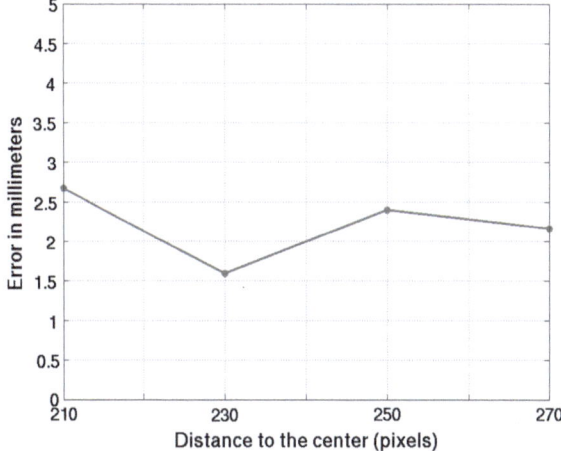

Fig. 3.29 3D reconstruction error in mm. The horizontal axis represents the different radii r

example can be observed in Fig. 3.28. We particularly observed that this approach is quite sensitive to the length and the position of the lines. We show the results where the calibration was possible in the corresponding radii. Figures 3.29, 3.30 show the 3D reconstruction error and the reprojection error, respectively. We observe a similar behavior to the other approaches, but having a bigger error, both in the 3D reconstruction error and the reprojection error.

One may think that a comparison of this method using just a few representative elements, in this case lines, present in one single image, against others where hundreds of representative elements (2D points), are extracted from several images, might be unfair. In this order we tried to calibrate the central catadioptric systems using as many lines as possible, present in the same images of the 3D pattern used to calibrate the system using the other approaches. A few examples of the lines used are shown in Fig. 3.31. The results calibrating the central catadioptric systems using this method

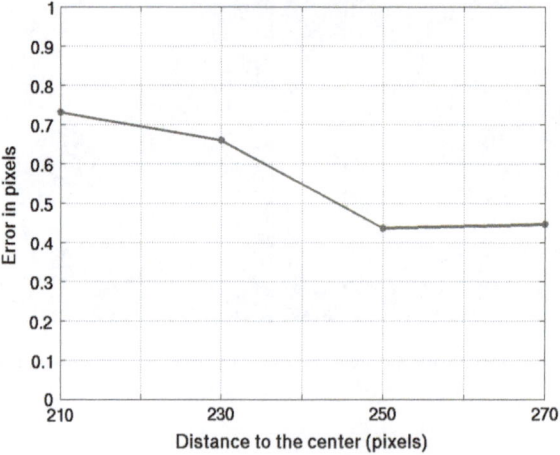

Fig. 3.30 Reprojection error in pixels. The horizontal axis represents the different radii r

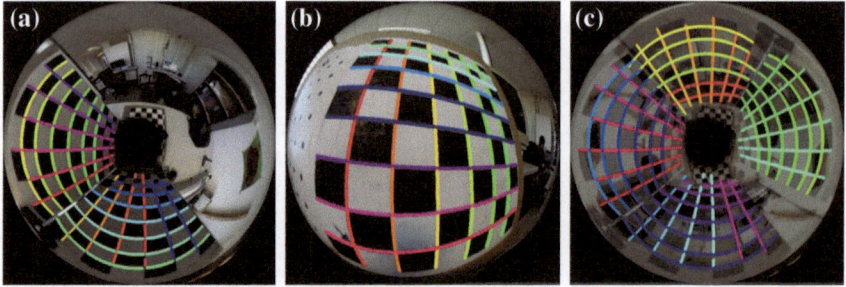

Fig. 3.31 Different configurations tested to calibrate the central catadioptric systems using the Sphere-Lines approach. **a** Unknown-shape system, **b** Fish-eye lens and **c** a combination of two images to cover the whole omnidirectional image with lines

did not succeed. We obtained several types of errors and sometimes convergence problems. Because of that we calibrate the central catadioptric systems using the maximum number of lines present in one single image, different from the ones used by the other approaches.

3.4.7 Discussion

After all these experiments with different systems we observe that all the approaches give good results, with the exception of the Sphere-Lines approach with the fish-eye system, basically because this approach is not designed to deal with such systems. In particular, for the fish-eye lens, the best calibration was achieved with the DLT-like approach. In the case of the unknown-shape camera, the Distortion-2D Pattern approach provided the best result. With respect to the noncentral hypercatadioptric, DLT-like, Sphere-2D Pattern and Distortion-2D Pattern approaches all gave similar accuracy. Finally, the hypercatadioptric system was calibrated slightly better by both the DLT-like and the Sphere-2D Pattern approaches. We also analyze the importance of the area occupied by the calibration elements (points, lines) in the calibration image(s). All approaches require this area to be as big as possible to compute a good calibration. In the particular case of the Sphere-Lines approach, the lines must be as large as possible and must intersect far from the image center. In terms of computing performance, all these approaches perform a nonlinear step after the initialization of the intrinsic and extrinsic parameters is computed. The DLT-like approach is the fastest since it estimates less parameters. Next are the Sphere-2D Pattern and the Distortion-2D Pattern with several extrinsic parameters corresponding to each image to compute, plus the intrinsic parameters. The slowest method is the Sphere-Lines approach, since it uses a complex geometry to compute the self-polar triangle and takes into account every single pixel contained in the line images.

We also consider the importance on what we need to make these methods to work and the problems observed at the calibration time.

- *Sphere-2D Pattern, Distortion-2D Pattern.* These two approaches require multiple images of a 2D pattern to perform the calibration. Both of them have automatic corner extractors but most of the times these do not work properly and the points have to be given manually. This is the most tedious part since we have a minimum of 8 to 10 images, each image containing 48 points giving a total of $384 \sim 480$ points. Besides that the Sphere-2D Pattern approach requires the user to indicate the image center and a minimum of three nonradial points to estimate the focal length.
- *DLT-like.* This approach does not require any prior information but one single omnidirectional image containing 3D points spread on three different planes. The inconvenient with this method is to obtain the 3D points contained in the 3D pattern. All the image points in the 3D pattern images are given manually. We observed that depending on the image used the mirror parameter ξ is better or worse estimated. Something similar happens with the Sphere-2D Pattern approach.
- *Sphere-Lines.* This approach requires the easiest setting to be constructed. It only requires one single omnidirectional image containing a minimum of three lines. One thing observed using this approach is that it strongly depends on the principal point estimation. If this estimation is not accurate enough the calibration is not performed properly. Also we observe some difficulties while calibrating the unknown shape catadioptric system. The number and the location of lines in the image is important to correctly calibrate the system. Sometimes using more than three lines we had convergence problems or we obtained calibrations containing non-real solutions.

Notice that each method has its own manner to extract the points from the images. In this order, we should decouple the matching process from the parameter estimation process.

3.5 Closure

In this chapter, we have presented a comparison of four methods to calibrate omni-directional cameras available as OpenSource. Two of them require images of a 2D pattern, one requires images of lines and the last one requires one image of a 3D pattern. Three of these approaches use the sphere camera model. This model can give some information about the mirror of the omnidirectional system besides it provides a theoretical projection function. The other approach is based on a distortion function. Both models can deal with any central catadioptric system and fish-eyes. However, the Sphere-Lines approach that uses the sphere camera model cannot deal with the fish-eye system. All these approaches use a nonlinear step which allows them to have a reprojection error less than 1 pixel. In this chapter, we perform a SfM experiment to compare the different approaches with useful criteria. This experiment showed that the calibration reached by any of these methods can give accurate reconstruction results. The distribution of the points in the omnidirectional images is important in order to have an accurate calibration. These points have to cover as much as possible of the omnidirectional image and mainly in the peripheric area.

Chapter 4
Two-View Relations Between Omnidirectional and Conventional Cameras

Abstract In this chapter, we present a deep analysis of the hybrid two-view relations combining images acquired with uncalibrated central catadioptric systems and conventional cameras. We consider both, hybrid fundamental matrices and hybrid planar homographies. These matrices contain useful geometric information. We study three different types of matrices, varying in complexity depending on their capacity to deal with a single or multiple types of central catadioptric systems. The first and simplest one is designed to deal with paracatadioptric systems, the second one and more complex, considers the combination of a perspective camera and any central catadioptric system. The last one is the complete and generic model which is able to deal with any combination of central catadioptric systems. We show that the generic and most complex model sometimes is not the best option when we deal with real images. Simpler models are not as accurate as the complete model in the ideal case, but they provide a better and more accurate behavior in presence of noise, being simpler and requiring less correspondences to be computed. Experiments with synthetic data and real images are performed. With the use of these approaches, we develop the successful hybrid matching between perspective images and hypercatadioptric images using SIFT descriptors.

4.1 Introduction

The combination of central catadioptric cameras with conventional ones is relevant since a single catadioptric view contains a more complete description of the scene, and the perspective image gives a more detailed description of the particular area or object we are interested in. Some areas where the combination of these cameras has an important role are localization and recognition (Menem and Pajdla 2004), since a database of omnidirectional images would be more representative with fewer points of view and less data, and perspective images are the simplest query images. In visual surveillance (Chen et al. 2003), catadioptric views provide coarse information about

L. Puig and J. J. Guerrero, *Omnidirectional Vision Systems*, SpringerBriefs
in Computer Science, DOI: 10.1007/978-1-4471-4947-7_4, © Luis Puig 2013

locations of the targets, while perspective cameras provide high resolution images for more precise analysis. In active vision (Jankovic and Naish 2007), this mixture is naturally implemented; an omnidirectional camera provides peripheral vision, while a controllable perspective camera provides foveal vision.

We are particularly interested in the two-view relations between uncalibrated catadioptric and conventional views working directly in the raw images. In the literature very few approaches are presented to compute hybrid two-view relations mixing uncalibrated catadioptric and conventional cameras. These approaches have in common the use of lifted coordinates to deal with the nonlinearities of the catadioptric projection. Most of these approaches are presented theoretically and with simple experiments. In this book, we tackle this situation by performing a deep evaluation of the hybrid fundamental matrix and the hybrid planar homography using simulated data and real images. In (Puig et al. 2012b), this analysis was presented.

To compute the two-view geometry, we require pairs of corresponding points between the views. These correspondences are built from previously detected relevant features. Perhaps the most used extractor is the SIFT (Lowe 2004). However, if SIFT features extracted in an omnidirectional image are matched to features extracted in a perspective image, the results are not good; this is because the SIFT descriptor is scale-invariant but not projective invariant. We observe that with a simple flip of the omnidirectional image, SIFT matching can still be useful, requiring neither image rectification nor panoramic transformation. Moreover, with the combination of a RANSAC approach with the hybrid two-view relations, we are able to perform the automatic robust matching between a conventional image and a catadioptric one without requiring unwarping or any other transformation of the catadioptric image.

4.2 Related Work

The multiview geometry problem for central catadioptric systems has been studied in recent years. Some approaches require the calibration of the systems. Svoboda and Pajdla (2002) study the epipolar geometry for central catadioptric systems. Based on the model of image formation, they propose the epipolar geometry for catadioptric systems using elliptic, hyperbolic, and parabolic mirrors. Geyer and Daniilidis (2001b, 2002b) have shown the existence of a fundamental matrix for paracatadioptric cameras. This has been extended by Sturm (2002) toward fundamental matrices and trifocal tensors for mixtures of paracatadioptric and perspective images. Barreto (2006) and Barreto and Daniilidis (2006) showed that the framework can also be extended to cameras with lens distortion due to the similarities between the paracatadioptric and division models. Recently, Sturm and Barreto (2008) extended these relations to the general catadioptric camera model, which is valid for all central catadioptric cameras. They showed that the projection of a 3D point can be modeled using a projection matrix of size 6×10. They also show the existence of a general fundamental matrix for pairs of omnidirectional images, of size 15×15 and plane homographies, also of size 15×15.

In last years, some works have been developed considering the hybrid epipolar geometry for different combinations of uncalibrated central projection systems, including the pin-hole model. To a lesser extent, homographies have also been studied using uncalibrated central systems. They establish a relation between the projections on the omnidirectional images of 3D points that lie on a plane. In a seminal work, Sturm (2002) proposes two models of hybrid fundamental matrices, a 4×3 fundamental matrix to relate a paracatadioptric view and a perspective view and a 6×3 fundamental matrix to relate a perspective view and a general central catadioptric view. He also describes the 3×4 plane homography which represents the mapping of an image point in a paracatadioptric view to the perspective view. This mapping is unique, unlike the opposite one that maps a point in the perspective image to two points in the paracatadioptric image. He also shows how to use the homographies and fundamental matrices to self-calibrate the paracatadioptric system. All these methods use only lifted coordinates of the points in the omnidirectional image. Menem and Pajdla (2004) propose an algebraic constraint on corresponding image points in a perspective image and a circular panorama. They use a lifting from 3- to 6-vector to describe Plücker coordinates of projected rays. Claus and Fitzgibbon (2005) propose the lifting of image points to 6-vectors to build a general purpose model for radial distortion in wide angle and catadioptric lenses. Chen and Yang (2005) use their particular geometric projection of a parabolic mirror. They define two homographies, one for each direction of the mapping. A 3×4 mapping from the paracatadioptric image to the perspective image, and a 3×6 one from the perspective view to the paracatadioptric, which because of its nonlinearity requires an iterative process to be computed. Barreto and Daniilidis (2006) propose a general model that relates any type of central cameras including catadioptric systems with mirrors and lenses and conventional cameras with radial distortion. They apply the lifted coordinates in both images. These lifted coordinates correspond to a map from $\wp^2$ to $\wp^5$ through Veronese maps, generating a 6×6 fundamental matrix.

Recently, Sturm and Barreto (2008) presented the general catadioptric fundamental matrix, a 15×15 matrix which uses a double lifting of the coordinates of the points in both images. In this work, they also present the general omnidirectional plane homography. It corresponds to a 15×15 matrix that relates the fourth order Veronese map of a point in the first omnidirectional or perspective image to a quartic curve in the corresponding second image. To compute the 225 elements of such a matrix, 45 correspondences of points are required since every correspondence gives five equations. Gasparini et al. (2009) extend this work by simplifying the catadioptric homography to a 6×6 matrix that relates a 6-vector that corresponds to the lifting of a 3D point lying on a plane and a degenerated dual conic also represented by a 6-vector.

4.3 Hybrid Two-View Relations of Uncalibrated Images

In this section, we explain the hybrid epipolar geometry between catadioptric and perspective images. We also explain the hybrid homography induced by a plane observed in these two types of image. For both two-view relations, we show and analyze three different models. F66, F36, and F34 for fundamental matrices and H66, H36, and H34 for hybrid homographies.

4.3.1 Hybrid Fundamental Matrices

Similar to the conventional fundamental matrix, the hybrid fundamental matrix encapsulates the intrinsic geometry between two views. In this case, a perspective view and a catadioptric one. It is independent of the observed 3D scene and it can be computed from correspondences of imaged scene points. This geometry can be observed in Fig. 4.1. Under this hybrid epipolar geometry, a point in the perspective image $\mathbf{q}_p$ is mapped to its corresponding epipolar conic in the catadioptric image $\mathbf{c}$

$$\mathbf{c} \sim \mathsf{F}_{cp}\mathbf{q}_p. \tag{4.1}$$

The particular shape of the epipolar conics depends on the central projection system we are dealing with, the angle between the epipolar plane and the mirror symmetry axis and the orientation of the perspective/orthographic camera with respect to the mirror. In the case of a paracatadioptric system (with a perfectly orthographic camera), the shape of the epipolar conic corresponds to a circle (Svoboda and Pajdla 2002) and can be represented as a 4-vector $\mathbf{c}_{par} = (c_1, c_2, c_3, c_4)^\mathsf{T}$. In the case of hypercatadioptric systems, the epipolar conics can be ellipses, hyperbolas, parabolas, or lines. They are represented as 6-vectors $\mathbf{c}_{hyp} = (c_1, c_2, c_3, c_4, c_5, c_6)^\mathsf{T}$. The representation of such conics as matrices is

Fig. 4.1 Hybrid epipolar geometry

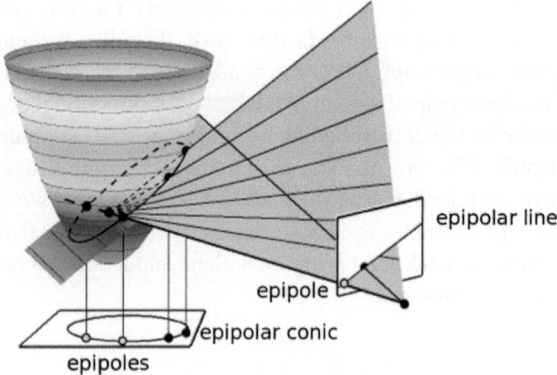

$$\Omega_{\text{par}} = \begin{pmatrix} 2c_1 & 0 & c_2 \\ 0 & 2c_1 & c_3 \\ c_2 & c_3 & 2c_4 \end{pmatrix}, \quad \Omega_{\text{hyp}} = \begin{pmatrix} 2c_1 & c_2 & c_4 \\ c_2 & 2c_3 & c_5 \\ c_4 & c_5 & 2c_6 \end{pmatrix}. \tag{4.2}$$

To verify if a point $\mathbf{q}_c = (q_1, q_2, q_3)^\top$ belongs to a certain conic, it must satisfy the identity $\mathbf{q}_c^\top \Omega \mathbf{q}_c = 0$. If we develop this identity with $\Omega = \{\Omega_{\text{par}}, \Omega_{\text{hyp}}\}$ we will obtain the expression $\hat{\mathbf{q}}_c \mathbf{c} = 0$ with $\mathbf{c} = \{\mathbf{c}_{\text{par}}, \mathbf{c}_{\text{hyp}}\}$ and $\hat{\mathbf{q}}_c$ the lifted coordinates we already studied in Sect. 1.3. In the first case as a 4-vector $\hat{\mathbf{q}}_c = (q_1^2 + q_2^2, q_1 q_3, q_2 q_3, q_3^2)^\top$ and as a 6-vector $\hat{\mathbf{q}}_c = (q_1^2, q_1 q_2, q_2^2 q_1 q_3, q_2 q_3, q_3^2)^\top$ in the second one.

In the opposite direction, from the catadioptric view to the perspective one, a point in the catadioptric image $\mathbf{q}_c$ has to be lifted $\hat{\mathbf{q}}_c$ and then mapped to its corresponding epipolar line in the perspective image

$$\mathbf{l} \sim \mathsf{F}_{cp}^\top \hat{\mathbf{q}}_c. \tag{4.3}$$

In general, the relation between catadioptric and perspective images with the fundamental matrix that we call *hybrid fundamental matrix* is established by

$$\mathbf{q}_p^\top \mathsf{F}_{pc} \hat{\mathbf{q}}_c = 0. \tag{4.4}$$

Using the different lifted coordinates (cf. Sect. 1.3) in either the catadioptric image or both the perspective and the catadioptric image, we can define different fundamental matrices.

4.3.1.1 General Catadioptric System, F66 and F36

As mentioned before, the generic fundamental matrix between two central projection systems (including catadioptric and perspective ones) is a 15×15 matrix, which uses a double lifting of the coordinates of the points in both the omnidirectional and the perspective image (Sturm and Barreto 2008). This lifting represents quartic epipolar curves. Since this matrix is intractable in a practical way, we prefer to refer to fundamental matrices easier to compute and that have been successfully applied.

Barreto and Daniilidis (2006) propose a 6×6 fundamental matrix, which is able to deal with different combinations of central catadioptric systems and conventional cameras. This matrix is obtained from the lifted coordinates $\hat{\mathbf{q}}_c$ of points in the omnidirectional and the lifted coordinates $\hat{\mathbf{q}}_p$ of points in the perspective images.

$$\hat{\mathbf{q}}_p^\top \mathsf{F}_{pc} \hat{\mathbf{q}}_c = 0 \tag{4.5}$$

This matrix establishes a bilinear relation between the two views, relating a point in the omnidirectional image to a conic in the perspective one $\mathbf{c}_p \sim \mathsf{F}_{pc} \hat{\mathbf{q}}_c$. This conic is composed by two lines. These lines are the forward looking epipolar line and the backward looking epipolar line. These lines can be extracted from an SVD of the epipolar conic. This 6×6 fundamental matrix is named F66 and corresponds to the theoretically complete model.

A simpler formulation is established by Sturm (2002) where the lifted coordinates are only applied to the point in the catadioptric image: $\hat{\mathbf{q}}_c = (q_1^2, q_1 q_2, q_2^2, q_1 q_3, q_2 q_3, q_3^2)^\mathsf{T}$. This matrix establishes a relation between a perspective or affine view and a general central catadioptric view. In Sturm (2002) is mentioned that this matrix just works in one direction, but experiments in Puig et al. (2008) show that it works in both directions. We name this 3×6 fundamental matrix, F36. This matrix is an approximation to F66.

4.3.1.2 Paracatadioptric System, F34

The paracatadioptric system is a particular catadioptric system composed by a parabolic mirror and an orthographic camera. In this case, the shape of the epipolar conic is a circle and we use the simplified lifting $\hat{\mathbf{q}}_c = (q_1^2 + q_2^2, q_1 q_3, q_2 q_3, q_3^2)^\mathsf{T}$ (1.12) in the coordinates of the points in the omnidirectional image. The coordinates of the points in the perspective image are normal homogeneous coordinates. We name this 3×4 fundamental matrix, F34.

4.3.1.3 Computation of the Hybrid Fundamental Matrix

We use a DLT-like approach Hartley and Zisserman (2000) to compute the hybrid fundamental matrix. It is explained as follows. Given n pairs of corresponding points $\mathbf{q}_c \leftrightarrow \mathbf{q}_p$, solve the Eqs. (4.4) or (4.5) to find F_{cp}. The solution is the least eigenvector of $\mathsf{A}^\mathsf{T}\mathsf{A}$, where A^T is the equation matrix

$$\mathsf{A}^\mathsf{T} = \begin{pmatrix} \mathbf{q}_{p_1} q_{c_{1_1}} & \cdots & \mathbf{q}_{p_1} q_{c_{1_m}} \\ \vdots & \ddots & \vdots \\ \mathbf{q}_{p_n} q_{c_{n_1}} & \cdots & \mathbf{q}_{p_n} q_{c_{n_m}} \end{pmatrix}. \tag{4.6}$$

The number of pairs of corresponding points n needed to compute the hybrid fundamental matrix depends on the number of elements of the fundamental matrix to be computed. Each pair of corresponding points gives one equation. Therefore 35, 17, and 11 correspondences are required at least to compute the F66, F36, and F34, respectively. It is recommended that $n \gg size(\mathsf{F}_{pc})$. The number of parameters required to compute these approaches is crucial if we take into account that in wide-baseline image pairs, the correspondences are difficult to obtain, and much more in images coming from different sensor types. In this case, the F34 has a clear advantage over the other two more complex approaches.

4.3.1.4 Rank 2 Constraint

The above fundamental matrices are, like for the purely perspective case, of rank 2. If the task we are interested in requires the epipoles, it is mandatory to ensure that the estimated fundamental matrix has rank 2. To impose the rank 2 constraint, we have tried two options. One is to enforce this constraint minimizing the Frobenius norm using SVD as explained in (Hartley and Zisserman 2000) which we call *direct imposition* (DI). The other option is to perform a nonlinear re-estimation process minimizing the distance from points in one image to their corresponding epipolar conic or line in the other one, using the Levenberg–Marquardt (LM) algorithm. To guarantee the rank 2, we use a matrix parameterization proposed by Bartoli and Sturm (2004) which is called the *orthonormal representation* of the fundamental matrix. This approach was originally applied to $O(3)$ matrices and we adapt it to F34, F36, and F66.

4.3.1.5 Computing the Epipoles

The process involved in the computation of the epipoles from the three tested hybrid fundamental matrices, F34, F36, and F66 is based on the computation of their corresponding null spaces.

The hybrid F34 matrix has a one-dimensional left null space and one right null vector. This one corresponds to the epipole in the perspective image. The two epipoles in the omnidirectional image are extracted from the left null space. We have to compute the left null vectors that are valid lifted coordinates. From these, the epipoles are extracted in a closed form. The 4-vectors must satisfy the following quadratic constraint:

$$\hat{\mathbf{q}}_c \sim \begin{pmatrix} q_1^2 + q_2^2 \\ q_1 q_3 \\ q_2 q_3 \\ q_3^2 \end{pmatrix} \Leftrightarrow \hat{q}_{c_1} \hat{q}_{c_4} - \hat{q}_{c_2}^2 - \hat{q}_{c_3}^2 = 0 \qquad (4.7)$$

In the case of the F36 matrix, we follow a similar process. The epipole in the perspective image is also given by the right null vector and the two epipoles of the omnidirectional image are extracted from the left null space of this matrix. In this case, the valid lifted coordinates have to satisfy the following equations:

$$\hat{\mathbf{q}}_c \sim \begin{pmatrix} q_1^2 \\ q_1 q_2 \\ q_2^2 \\ q_1 q_3 \\ q_2 q_3 \\ q_3^2 \end{pmatrix} \Leftrightarrow \begin{aligned} \hat{q}_{c_1} \hat{q}_{c_3} - \hat{q}_{c_2}^2 &= 0 \\ \hat{q}_{c_1} \hat{q}_{c_6} - \hat{q}_{c_4}^2 &= 0 \\ \hat{q}_{c_2} \hat{q}_{c_6} - \hat{q}_{c_5}^2 &= 0 \end{aligned} \qquad (4.8)$$

In the case of the F66 matrix, we use the same approach as for F36 to extract the epipoles corresponding to the omnidirectional image from the left null space of the fundamental matrix. For the epipole in the perspective one, a different process is required. We extract it from the null vector of the degenerate epipolar conic $\Omega_p \sim$ F66$\hat{q}_c$ projected from a point in the omnidirectional image to the perspective image. This conic contains the two points q_+ and q_-.

4.3.2 Hybrid Homography

Hybrid homographies relate the projections of points that lie on a plane on different types of images. In particular, we analyze the homographies that relate omnidirectional and perspective images. Similarly as before with fundamental matrices, we consider three different models. The general model H66 and two simplified models H36 and H34.

4.3.2.1 Generic Model, H66

From (Sturm and Barreto 2008) the projection of a 3D point in any central catadioptric system using lifted coordinates can be described by a 6×10 projection matrix P_{cata}

$$\hat{q} \sim P_{cata}\widehat{Q}, \quad P_{cata} = \widehat{K}X_\xi \widehat{R}_{6\times 6} \left(I_6 \; T_{6\times 4} \right) \tag{4.9}$$

If we assume that the 3D points lie on a plane $z = 0$, $Q = (Q_1, Q_2, 0, 1)^\mathsf{T}$, the non-zero elements of its lifted representation is a 6-vector $\widehat{Q}_c = (Q_1^2, Q_1 Q_2, Q_2^2, Q_1, Q_2, 1)^\mathsf{T}$ and the projection matrix reduces to size 6×6:

$$H66 = \widehat{K}X_\xi \widehat{R}\,(I_{6\times 3}[t_1 t_2 t_4]) \tag{4.10}$$

where t_i is the ith column of the matrix T and H66 is the 6×6 homography matrix relating the lifting of the 2D coordinates of the points on the plane to their dual conic representation on the image plane Ω as explained in Sect. 1.2.

This homography can also relate the projection of these 3D points in two different images. In particular, in two images acquired with different sensors, a conventional one q_p and an omnidirectional one q_c.

$$\hat{q}_p \sim H66\,\hat{q}_c \tag{4.11}$$

To compute this homography, we use a DLT-like (Direct Linear Transformation) approach. As in the perspective case, we need correspondences $q_p^i \leftrightarrow q_c^i$ between points lying on the plane in the conventional image q_p^i and in the omnidirectional one q_c^i. From (4.11) we obtain

$$\widehat{[\mathbf{q}_p]}_\times \mathsf{H66}\, \hat{\mathbf{q}}_c = \mathbf{0} \tag{4.12}$$

If the jth row of the matrix H66 is denoted by $\mathbf{h}_j^\mathsf{T}$ and arranging (4.12) we have

$$\widehat{[\mathbf{q}_p]}_\times \otimes \hat{\mathbf{q}}_c \begin{pmatrix} \mathbf{h}_1^\mathsf{T} \\ \mathbf{h}_2^\mathsf{T} \\ \mathbf{h}_3^\mathsf{T} \\ \mathbf{h}_4^\mathsf{T} \\ \mathbf{h}_5^\mathsf{T} \\ \mathbf{h}_6^\mathsf{T} \end{pmatrix} = \mathbf{0} \tag{4.13}$$

These equations have the form $\mathsf{A}^i\mathbf{h} = \mathbf{0}$, where A^i is a 36×6 matrix, and $\mathbf{h} = (\mathbf{h}_1^\mathsf{T}, \mathbf{h}_2^\mathsf{T}, \mathbf{h}_3^\mathsf{T}, \mathbf{h}_4^\mathsf{T}, \mathbf{h}_5^\mathsf{T}, \mathbf{h}_6^\mathsf{T})^\mathsf{T}$ is a 36-vector made up of the entries of matrix H66. The matrix A^i has the form

$$\mathsf{A}^i = \begin{pmatrix} 0 & 0 & q_3^2\hat{\mathbf{q}}_c & 0 & -2q_3q_2\hat{\mathbf{q}}_c & q_2^2\hat{\mathbf{q}}_c \\ 0 & -q_3^2\hat{\mathbf{q}}_c & 0 & q_3q_2\hat{\mathbf{q}}_c & q_3q_1\hat{\mathbf{q}}_c & -q_2q_1\hat{\mathbf{q}}_c \\ q_3^2\hat{\mathbf{q}}_c & 0 & 0 & -2q_3q_1\hat{\mathbf{q}}_c & 0 & q_1^2\hat{\mathbf{q}}_c \\ 0 & q_3q_2\hat{\mathbf{q}}_c & -q_3q_1\hat{\mathbf{q}}_c & -q_2^2\hat{\mathbf{q}}_c & q_2q_1\hat{\mathbf{q}}_c & 0 \\ -q_3q_2\hat{\mathbf{q}}_c & q_3q_1\hat{\mathbf{q}}_c & 0 & q_2q_1\hat{\mathbf{q}}_c & -q_1^2\hat{\mathbf{q}}_c & 0 \\ q_2^2\hat{\mathbf{q}}_c & -2q_2q_1\hat{\mathbf{q}}_c & q_1^2\hat{\mathbf{q}}_c & 0 & 0 & 0 \end{pmatrix} \tag{4.14}$$

and is rank 3, so each correspondence gives three independent equations. Thus, we need at least 12 correspondences to compute H66 (Gasparini et al. 2009).

4.3.2.2 Simplified Homographies H34 and H36

We also consider two approximations of the hybrid homography. H34 and H36 are the two hybrid homographies that map a lifted vector (1.11) or (1.12) corresponding to a point in the omnidirectional image $\hat{\mathbf{q}}_c$ to a point in the corresponding plane $\mathbf{q}_p$ in homogeneous coordinates. The former is related to the theoretical model of a paracatadioptric system and the latter considers any central catadioptric system. Similar to (4.13), we consider the Kronecker product of $\mathbf{q}_p$ and $\hat{\mathbf{q}}_c$. Both homographies are computed using a DLT approach. Since each correspondence gives two equations, we require at least six correspondences to compute H34 and nine correspondences to compute H36.

4.4 Evaluation of the Hybrid Two-View Models

In this section, we analyze the behavior of the three fundamental matrices (F66, F36, and F34) and the three homographies (H66, H36, and H34). We present some experiments performed with synthetic data and real images.

4.4.1 Simulated Data

We use a simulator which generates omnidirectional images coming from a hyper-catadioptric system and perspective images from a pin-hole model. The two sensors are placed in a virtual volume of $5 \times 2.5 \times 7$ m. width, height, and depth, respectively, where points are located randomly ($n \gg 35$) in the case of the fundamental matrix and in planes in the case of the homographies. The perspective camera has a resolution of 1000×1000 pixels and is located at the origin of the coordinate system. The omnidirectional camera is located to have a good view of the whole scene. We use the sphere camera model (Barreto and Daniilidis 2006) to generate the omnidirectional image. We consider two real hypercatadioptric systems with mirror parameters of $\xi = 0.9662$ (m1) and $\xi = 0.7054$ (m2), from two real hyperbolic mirrors designed by Neovision[1] and Accowle,[2] respectively. As a common practice and because we are using lifted coordinates, we apply a normalization to the image coordinates where the origin is the image center and the width and height are 1. Once the points are projected we add Gaussian noise, described by σ, in both images.

4.4.1.1 Analysis of the Fundamental Matrices

The fundamental matrices are computed using a Levenberg–Marquardt[3] nonlinear minimization of the geometric distance from image points to epipolar lines and conics using the point to conic distance proposed by Sturm and Gargallo (2007). For every σ representing the amount of noise, we repeat the experiment 10 times to avoid particular cases due to random noise. We show the mean of these iterations. Figure 4.2 shows the distances from points to their corresponding epipolar conics and lines as a function of image noise.

From Fig. 4.2 we can observe that when there is no noise present in the image, the F66 shows the best performance, which is expected since F66 is the theoretically correct model. This changes when noise increases. In this case, F34 and F36 show a better performance, being consistent with the noise present in the images. The F34 shows a better performance with the mirror m1 since this one is closer to a parabolic

---

[1] http://www.neovision.cz

[2] http://www.accowle.com

[3] **lsqnonlin** function provided by Matlab.

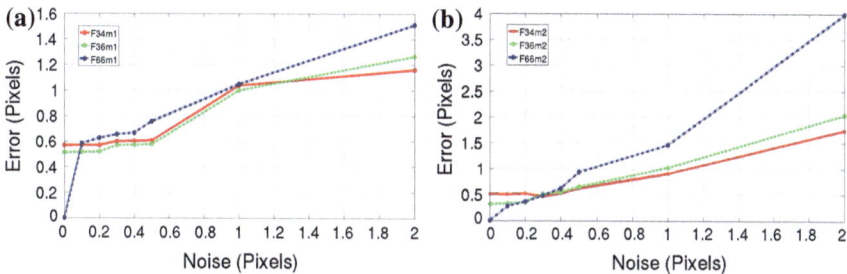

Fig. 4.2 Behavior of the three fundamental matrices in function of image noise (σ): RMSE of points to epipolar conics and lines using mirrors **a** m1 and **b** m2

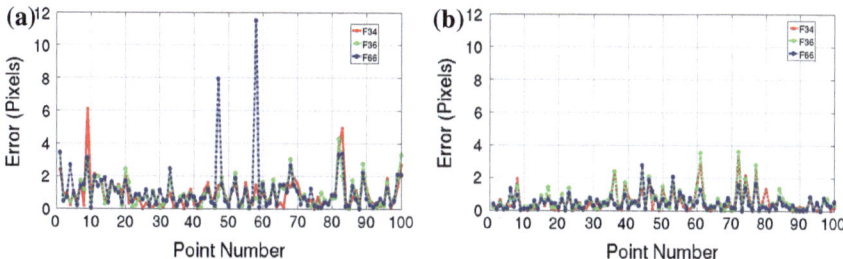

Fig. 4.3 RMSE error from points to their corresponding epipolar conic and lines. **a** Using the direct imposition of the rank 2. **b** Using the LM algorithm

mirror, the one the matrix F34 was designed to deal with. The residuals of F36 are slightly larger than the ones from F34. We observe that F66 is instable when noise is present in the images. This behavior can be caused by the over-parameterization of the model; the more the parameters, the higher the sensitivity to noise; it can also explain the difference between the F36 and F34.

We also estimate the epipoles from the three hybrid fundamental matrices, using the m1 hypercatadioptric system. In this experiment, we add $\sigma = 1$ pixel Gaussian noise to both images. We test the two approaches DI and LM to get a rank 2 matrix (cf. Sect. 4.3.1.4). We evaluate the performance of these approaches by the accuracy of the estimated epipoles and by the residual, which is the RMSE of the distances from the points used to compute the fundamental matrix to their corresponding epipolar lines and conics. In Fig. 4.3, we show the residuals for the three approaches imposing the rank 2 constraint by the *direct imposition* and by using the LM algorithm with *orthonormal representation*.

In Fig. 4.3a we can observe that F66 is very sensitive to the direct imposition of the rank 2 property with maximum errors of 8 and 12 pixels. This occurs because we are transforming a good solution that passes through the points in the perspective and omnidirectional images into a new matrix of rank 2 that contains the epipoles and makes all epipolar lines and conics to pass through them but far from the points in the corresponding images. This does not occur with the LM algorithm which uses the *orthonormal representation* because it imposes the rank 2 property and at the same time minimizes the distance between points and epipolar lines and conics, having a maximum error of 3 pixels.

Table 4.1 shows the epipoles from these two approaches. We can see from it that the three approaches give similar results in computing the epipole, but we also observe an increment in the distance from points to conics and the minimization obtained with the LM algorithm, all this as expected. Once more, F34 shows an interesting behavior giving a small distance to conics even with the DI approach. This adds another advantage to F34.

As observed from the previous experiments, F34 shows a good performance dealing with images coming from a hypercatadioptric system. In order to test this behavior, we designed the following experiment. We modify the mirror parameter ξ from the hyperbolic case ($0 < \xi < 1$) to the parabolic case ($\xi = 1$) (Barreto and Daniilidis 2006). We add $\sigma = 1$ pixels Gaussian noise in both images and repeat the experiment 10 times to avoid bias since we are using random noise. In Fig. 4.4, we observe that F34 can deal better with hypercatadioptric images when the mirror shape is close to a parabola ($\xi = 1$) and not as good as F66 and F36 models which are designed to deal with this type of systems but still having a RMSE of 1.28 pixels with the hyperbolic mirror defined by $\xi = 0.75$.

4.4.1.2 Analysis of Homographies

We perform experiments computing the hybrid homographies relating a plane in the ground and its projection in an omnidirectional image as well as the projection of a planar scene in both omnidirectional and perspective images. We use the same simulator as in the fundamental matrix case also considering the two different hyperbolic mirrors (m1) and (m2). The 3D points are distributed in a planar pattern. This pattern is composed of a squared plane with 11×11 points and a distance between points of 40 cm. The goal of the first experiment is to know the behavior of the three homography approaches in the presence of noise. We add different amounts of Gaussian noise described by its standard deviation (σ) to the coordinates of the points in the omnidirectional image. The DLT algorithm followed by a nonlinear step using LM minimizing errors in the image is used to compute the homographies. For every σ, we repeat the experiment 10 times in order to avoid particular cases due to random noise. The error of the projected points in the ground plane is shown in Fig. 4.5. We observe that the three approaches have a similar behavior. When the amount of noise is low, the best performance is given by H66, in fact it is the only one that has a zero error when we use noiseless data. When the amount of noise increases, the performance of H66 decreases and H34 and H36 remain with smaller errors. This result shows that H66 is more sensitive to noise than the other two approaches. The difference between the errors using the different mirrors is explained because the area occupied by the plane using m2 is bigger than the area covered using m1. With the m1 mirror we have errors of 5.2 cm with the H66, but with the m2 mirror this error decreases to 3 mm in both cases with $\sigma = 1$ pixel.

Table 4.1 Epipoles estimated by the three fundamental matrices

	True value	F66		F36		F34	
		DI	LM	DI	LM	DI	LM
e_1	(500, 303.07)	(499.24, 302.48)	(499.60, 303.42)	(500.23, 303.73)	(499.66, 303.98)	(500.07, 303.41)	(499.66, 303.95)
e_2	(500, 200)	(503.31, 199.17)	(501.55, 201.08)	(501.03, 201.27)	(501.52, 202.03)	(500.53, 201.79)	(501.29, 202.42)
RMSE	0.0	18.04	0.76	1.16	1.01	0.85	0.99

DI direct imposition, *LM* Levenberg-Marquardt

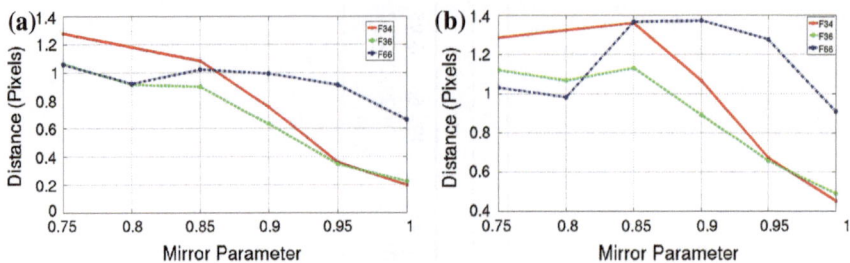

Fig. 4.4 Behavior of the three fundamental matrices as a function of the mirror parameter (ξ): mean distances from points to epipolar conics in **a** omnidirectional image, and **b** perspective image

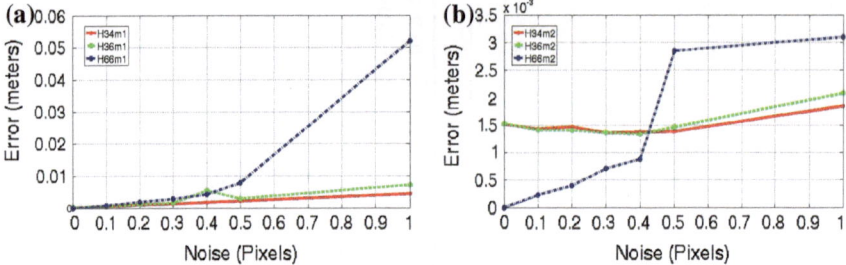

Fig. 4.5 Comparison between the three approaches to compute the hybrid homography. Using mirrors **a** m1, **b** m2

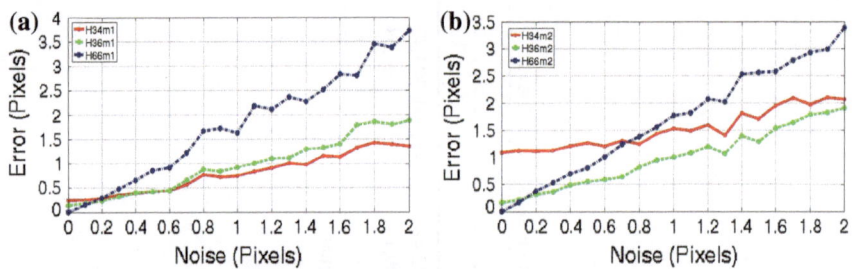

Fig. 4.6 Noise sensitivity of the hybrid homographies between omnidirectional and perspective images. **a** Using m1 and **b** using m2

The next experiment maps a point from a plane in a perspective image to its projection in the omnidirectional image. In this case, we added Gaussian to both perspective and omnidirectional image coordinates. We project a point from the omnidirectional to the perspective image, where the map is direct. In Fig. 4.6, we can observe the experiment using different amounts of Gaussian noise σ. Again H34 and H66 give better results than H66 except for the case with a very small amount of noise.

In the opposite direction, the homography maps a point in the perspective image to a conic in the omnidirectional one. Since the extraction of the corresponding point from this conic is difficult, a way to overcome this problem is to compute a different homography, which maps lifted coordinates in the perspective image to a single point in the omnidirectional one.

From the simulations, we observe that the hybrid fundamental matrices and the hybrid homographies with less parameters, F34 and H34 have a good performance even dealing with hyperbolic mirrors. Also they are less sensitive to noise than the theoretically correct and more general models F66 and H66. Note also that simpler models F34 and H34 require fewer point correspondences to be computed and therefore they have advantages in practice.

4.4.2 Experiments with Real Images

We also performed experiments with real images coming from a hypercatadioptric system and from a conventional camera. We compare the accuracy of the three methods to compute both the hybrid fundamental matrix and the hybrid homography.

4.4.2.1 Hybrid Fundamental Matrix

In this case, we use 70 manually selected pairs of corresponding points to compute the three approaches (F34, F36, and F66). In order to measure the performance of F, we calculate the root mean square error of the geometric distance from each correspondence to its corresponding epipolar conic or line. Table 4.2 shows these distances for the estimated F without imposing rank 2 and for the two ways to obtain the rank 2 fundamental matrix. We can observe that when we impose the rank 2, the error increases in particular with F66. With the *orthogonal normalization* using the LM algorithm F66 gives the best result but with very few difference with alternate models F34 and F36. When we impose the rank 2 constraint, we eliminate a few degrees of freedom of the matrix that better adjusts to the data, so the residual error must be worse actually. From Fig. 4.7 we can observe the epipolar lines and conics

Table 4.2 Mean of the distances to epipolar conics (D2C) and lines (D2L) for the 70 corresponding points in real images

	D2C			D2L		
	F34	F36	F66	F34	F36	F66
No rank 2	0.68	0.66	0.67	1.10	1.05	0.9
DI	0.87	1.64	21.06	1.36	2.07	3.82
LM	0.71	0.70	0.69	1.13	1.07	1.01

Using Direct Imposition (DI) and Levenberg-Marquardt (LM)

Fig. 4.7 **a–c** Epipolar conics using F34, F36, and F66. **d–f** *Epipolar lines* using F34, F36 and F66

from the three approaches. We also observed that a great number of correspondences, larger than the minimum are required to have a reasonable accuracy. Using F36, we obtain good results with 50 (three times the minimum) correspondences. This gives a good reason to use the F34 for further applications.

4.4.2.2 Hybrid Homographies

In this experiment, we select 55 correspondences manually. From these correspondences, we use 35 to compute the hybrid homographies H34, H36, and H66. We use the rest as test points. If we want to map points in the opposite direction, i.e., from the perspective image to the omnidirectional one, we require the inverse mapping of matrices H34 and H36. Since these matrices are not square, their computation is not possible. In this order, we compute separate homographies to map points in this direction. With this computation, we also avoid the extraction of the points from the corresponding conics. With respect to H66 it was shown in Gasparini et al. (2009) that two homographies have to be computed, since the inverse matrix does not correspond to the opposite mapping. In Fig. 4.8 we show the images used to compute the hybrid homographies. In Fig. 4.9a we show the error corresponding to the Euclidean distance between the estimated and the test points in the omnidirectional image. Figure 4.9b shows the error in the perspective image. We observe that H34 and H36 have a similar behavior. In both images, we also show the corresponding means of the error. H34 has the best performance in the perspective image, while H36 has it in the omnidirectional one. The worst behavior in both images corresponds to H66. All the approaches show a considerable error, which can be caused by the small area occupied by the points in the omnidirectional image.

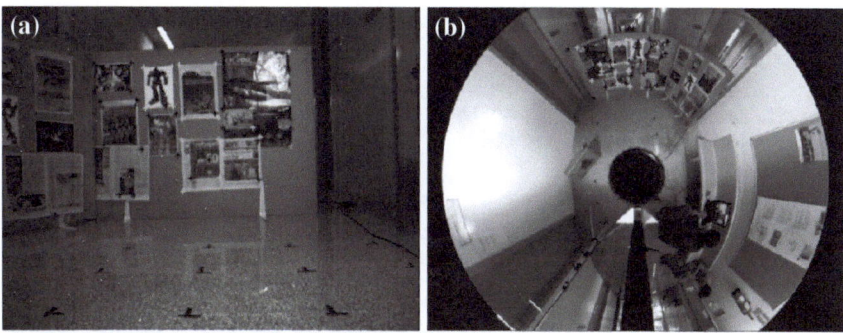

Fig. 4.8 Images used to compute the hybrid homographies. **a** Perspective image. **b** Omnidirectional image. Points used to compute the homographies are in *red* and test points in *green*

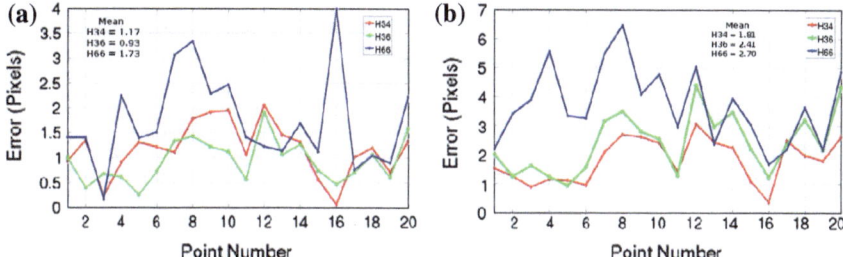

Fig. 4.9 Euclidean distance from the estimated points to the test points. **a** Omnidirectional image. **b** Perspective image

4.5 Automatic Matching Using Hybrid Two-View Relations

In this section, we present the robust automatic matching between uncalibrated hyper-catadioptric images and perspective images. The first step of a matching process between wide-baseline images is to obtain an initial or putative set of pairs of corresponding features. Reasonable matching of two omnidirectional images using well-known features like SIFT, SURF, or MSER has been reported (Murillo et al. 2007; Guerrero et al. 2008; Mauthner et al. 2006). It is known that the SIFT descriptor is scale invariant but not camera invariant, making it difficult to directly match omnidirectional images with perspective images using standard SIFT features. In Puig et al. (2008) we observed that by unwarping the omnidirectional image, we improved the number of matches between hybrid image pairs. This unwarping included a vertical flip of the omnidirectional image. In Puig and Guerrero (2009), we realized that this improvement was mainly caused by the simple flip of the omnidirectional image and not by the whole unwarping process. The SIFT descriptor is designed to be scale and rotation invariant and even camera invariant, if we consider that flipping an omnidirectional image can produce good matches with a conventional image.

Fig. 4.10 Matching directly the SIFT points in the omnidirectional and perspective images. **a** Using the unwarped image. **b** Using the flipped omnidirectional image. The matches with the normal omnidirectional image are not shown since near by all are outliers

We observe that the SIFT descriptor is not projective invariant, since the projective mirror effect is responsible for the majority of matching failures.

To evaluate that, we show in Fig. 4.10 the direct matching between SIFT points from a normal omnidirectional image and a perspective image. In this work, we use the SIFT implementation by Vedaldi (2007). The inliers and outliers obtained were counted manually. Table 4.3 shows that near by all matches are wrong if the omnidirectional image is directly used. Using the unwarped and the flipped transformation of the omnidirectional image, we repeat the experiment. In these cases, we observe an important increment on the number of correct matches showing both similar results. More results are shown in Table 4.4.

Note that this initial matching between the perspective and the flipped omnidirectional image has a considerable amount of inliers but also many outliers. This scenario requires a robust estimation technique and a geometric model to detect the

Table 4.3 Output from the SIFT matching using the original, unwarped, and flipped omnidirectional image

	SIFT points	Matches/inliers
Omnidirectional	2877	137/9
Unwarped image	4182	179/68
Flipped image	2867	207/76

Table 4.4 Numerical results of the hybrid matching using the set of images

	Unwarped omni SIFT	Flipped omni SIFT	Persp SIFT	Initial matches (inliers/outliers)		Robust epipolar geometry matches (inliers/outliers)	
				Unwarped	Flipped	Unwarped	Flipped
Experiment 1	3251	2867	1735	68/111	76/131	40/8	57/4
Experiment 2	4168	4172	1528	18/68	21/71	16/7	17/7
Experiment 3	3280	2967	1682	41/101	33/112	27/9	20/9
Experiment 4	2275	2208	15658	125/322	164/360	80/5	129/22

inliers and reject the outliers. Depending on the situation, either the hybrid epipolar geometry or the hybrid homography can be used.

4.5.1 Hybrid Fundamental Matrix

In a general case where the points are in any part of a 3D scene, the fundamental matrix is used. The automatic process to perform the matching between an omnidirectional image and a perspective one, using the hybrid fundamental matrix as geometric constraint is as follows:

1. **Initial Matching**. Scale-invariant features (SIFT) are extracted from perspective and flipped omnidirectional images and matched based on their intensity neighborhood.
2. **RANSAC robust estimation**. Repeat for r samples, where r is determined adaptively:

   a. Select a random sample of k corresponding points, where k depends on what model we are using (if $F34$, $k = 11$, if $F36$, $k = 17$ or if $F66$ $k = 35$). Compute the hybrid fundamental matrix F_{cp} as mentioned before.
   b. Compute the distance d for each putative correspondence, d is the geometric distance from a point to its corresponding epipolar conic (Sturm and Gargallo 2007).
   c. Compute the number of inliers consistent with F_{cp} by the number of correspondences for which $d < t$ pixels, t being a defined threshold.

Choose the F_{cp} with the largest number of inliers.
3. **Nonlinear re-estimation**. Re-estimate F_{cp} from all correspondences classified as inliers by minimizing the distance in both images to epipolar conics (Sturm and Gargallo 2007) and epipolar lines, using a nonlinear optimization process.

For the next experiments, we have selected the F34 model since its performance is similar to the other models and the number of correspondences required to be computed is the smallest. In a RANSAC approach, the number of parameters to estimate is important since it determines the number of iterations required. In practice, there is an agreement between the computational cost of the search in the space of solutions, and the probability of failure $(1 - p)$. A random selection of r samples of k matches ends up with a good solution if all the matches are correct in at least one of the subsets. Assuming a ratio ε of outliers, the number of samples to explore is $r = \frac{\log(1-p)}{\log(1-(1-\varepsilon)^k)}$. For example using a probability $p = 99\%$ of not failing in the random search and 30 % of outliers (ε), 231 iterations are needed to get a result using the F34. If we use the F36, 1978 iterations are needed for the same level of confidence. In the case of the F66, the number of iterations increases to 1.2×10^6 and becomes prohibitive for some applications.

Several omnidirectional and perspective image pairs are used to perform the experiment of automatic matching (Fig. 4.11). We avoid the rank 2 constraint since we are just concerned about the matching problem. Table 4.4 summarizes the results giving the number of SIFT features extracted in the two valid versions of omnidirectional images tested (unwarped and flipped), and in the perspective one. It also shows the quantity of inliers and outliers in the initial (SIFT matching) and the robust matching (hybrid fundamental matrix), using both the unwarped and the flipped transformations. Figure 4.12 shows two examples of the matching between omnidirectional and perspective images. In Experiment 1, we use images Fig. 4.11a, e. The number of SIFT points extracted from the flipped and the unwarped images are similar. We

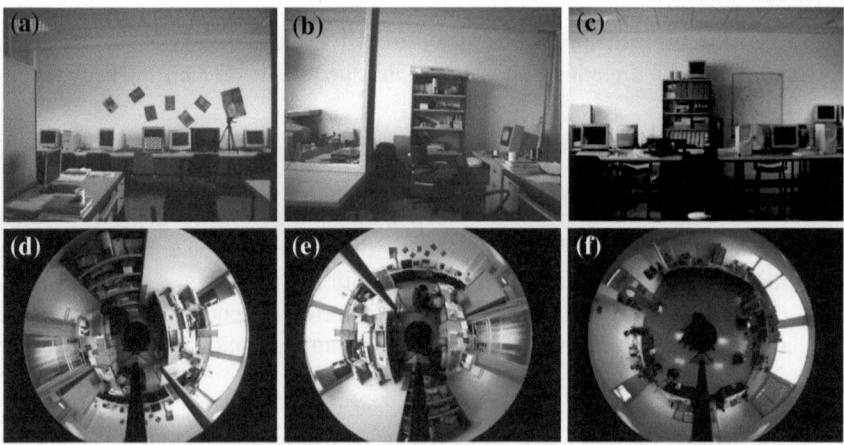

Fig. 4.11 Some of the images used to test the automatic matching using the fundamental matrix

Fig. 4.12 Matching between omnidirectional and perspective image using the hybrid epipolar geometry. **a** Experiment 1. **b** Experiment 4

observe that the initial matches are similar using the unwarped and flipped versions of the omnidirectional image, with a small advantage for the flipped one. These results confirm that SIFT is not projective invariant but it works well with such a distortion of catadioptric cameras. Despite the use of either of the transformed omnidirectional images, the automatic matching using the hybrid epipolar geometry is able to filter most of the outliers. In Experiment 4, the increment in the SIFT features of the perspective image is caused by the resolution of the image (1280 × 960). The results show that the hybrid epipolar constraint eliminates most of the outliers.

4.5.2 Hybrid Homography

Analogous to the matching process using the hybrid fundamental matrix we can use the hybrid homography when most of the scene points lie in a plane. An example of the automatic matching using the hybrid homography as a geometrical constraint can be observed in Fig. 4.13. Figure 4.13a shows the putative correspondences given by the SIFT matching and Fig. 4.13b after applying the robust matching process, computing a 3 × 4 homography.

Fig. 4.13 Matching between omnidirectional and perspective image using, **a** putative matches, **b** matches after the robust estimation using the hybrid homography

4.6 Closure

In this work we have presented a deep analysis of the two-view geometry combining a central catadioptric system and a conventional camera. In particular, we studied the hybrid epipolar geometry and the hybrid planar homography. We use lifted coordinates to generalize the two-view constraints, well-known for perspective image pairs. We selected three approaches to compute the hybrid fundamental matrix F34, F36, and F66 and three approaches to compute the hybrid homography H34, H36, and H66. We performed several experiments comparing the different approaches from the more complex and complete (F66, H66) to a more particular and simplified one (F34, H34), that in principle only can deal with a certain type of central catadioptric systems. From the simulation and real data experiments, these simplified models obtained better results in presence of noise. We observed that the complete models can deal with any catadioptric system under ideal conditions, but these approaches are more sensitive to the presence of noise. We successfully introduce the geometrical constraints in a robust matching process with initial putative matches given by SIFT points computed in the perspective image and the flipped version of the catadioptric one.

Chapter 5
Generic Scale-Space for a Camera Invariant Feature Extractor

Abstract In this chapter, we present a new approach to compute the scale-space of any omnidirectional image acquired with a central projection system. When these cameras are central they are explained using the sphere camera model, which unifies in a single model, conventional, paracatadioptric, and hypercatadioptric systems. Scale space is essential in the detection and matching of interest points, in particular scale-invariant points based on Laplacian of Gaussians, like the well known SIFT. We combine the sphere camera model and the partial differential equations framework on manifolds, to compute the Laplace–Beltrami (LB) operator which is a second order differential operator required to perform the Gaussian smoothing on catadioptric images. We perform experiments with synthetic and real images to validate the generalization of our approach to any central projection system.

5.1 Introduction

Image processing has developed through the years techniques for conventional (perspective) cameras. Among all these techniques, feature detection/extraction is one of the most relevant, since it represents a crucial step on higher level techniques, such as matching, recognition, structure from motion, SLAM, navigation, visual localization, visual control, surveillance, and many more. A particular useful property for features is to be scale-invariant. There are different approaches to detect scale-invariant features. Some of them are based on the scale-space analysis (Lowe 2004; Mikolajczyk and Schmid 2004). Some others are based on the gray scale intensity (Kadir and Brady 2001; Matas et al. 2002). SIFT Lowe (2004) has become the most used feature extraction approach. It has also been used directly in omnidirectional images (Guerrero et al. 2008), although it is not designed to work on them. This SIFT approach has inspired different works trying to replicate its good results on different imagery systems, in particular on wide-angle cameras. In Bulow (2004) a Gaussian kernel is derived. It requires the omnidirectional image to be mapped to the sphere. Then, the spherical Fourier transform is computed and convolved with the spherical Gaussian function. In Hansen et al. (2007) the image is mapped to the sphere and obtain scale-space images as the solution to the heat diffusion equation on the sphere which is implemented in the frequency domain using spherical harmonics. This approach introduces new drawbacks as aliasing and bandwidth selection. A complete SIFT version computed on the sphere also using the heat diffusion equation is presented by Cruz et al. (2009). In Hansen et al. (2010) an approximation to spherical diffusion using stereographic projection is proposed. It maps the omnidirectional image to the stereographic image plane through the sphere. It also maps the spherical Gaussian function to an equivalent kernel on the stereographic image plane. Then, the approximate spherical diffusion is defined as the convolution of the stereographic image with the stereographic version of the Gaussian kernel. More recently Lourenço et al. (2010) proposed an improvement to the SIFT detector by introducing radial distortion into the scale-space computation. In Bogdanova et al. (2007) a framework to perform scale-space analysis for omnidirectional images using partial differential equations is proposed. It leads to the implementation of the linear heat flow on manifolds through the Laplace–Beltrami (LB) operator. Omnidirectional images are treated as scalar fields on parametric manifolds. Based on this work Arican and Frossard (2010) proposed a scale-invariant feature detector for omnidirectional images. They deal with the paracatadioptric projection which is equivalent to the inverse of the stereographic projection. They model this projection on the sphere and obtain its corresponding metric. This metric is conformal equivalent to the Euclidean one making the computation of the LB operator straightforward. Although this approach could work with any catadioptric system, the metric describing the reflecting surface (mirror) has to be provided, which in some cases can be difficult to obtain.

In this chapter we show a new approach to compute the scale-space for any central catadioptric system, which was presented in Puig and Guerrero (2011). We integrate the sphere camera model (Geyer and Daniilidis 2000; Barreto and Araujo 2001), which describes any central catadioptric system, selecting it by one single parameter, with the partial differential equations on manifolds framework through the heat diffusion equation (Bogdanova et al. 2007; Arican and Frossard 2010). Using this framework and the mirror parameter we compute the metric representing that particular reflecting surface. Then we use this metric to calculate the corresponding LB operator. This second-order operator allows us to perform the Gaussian smoothing on omnidirectional images.

5.2 Scale-Space for Central Projection Systems

In this section we integrate the sphere camera model (Sect. 1.2) that models any central projection system and the techniques developed to compute the differential operators on non-Euclidean manifolds (Bertalmío et al. 2001) such as the mirror surfaces present in catadioptric systems.

5.2.1 Differential Operators on Riemannian Manifolds

The scale-space representation $I(x, y, t)$ is computed using the heat diffusion equation and differential operators on the non-Euclidean manifolds. It is defined as

$$\frac{\partial I(x, y, t)}{\partial t} = \Delta I(x, y, t) \tag{5.1}$$

where Δ is the LB operator and t is the scale level. The initial condition is $I(x, y, t_0) = I(x, y)$ where $I(x, y)$ is the original image.

We briefly define the differential operators on the manifolds which make possible the computation of the LB operator. Let $\mathbb{M}$ be a parametric surface on $\mathbb{R}^3$ with an induced Riemannian metric g_{ij} that encodes the geometrical properties of the manifold.

In a local system of coordinates x^i on $\mathbb{M}$, the components of the gradient reads

$$\nabla^i = g^{ij} \frac{\partial}{\partial x^j} \tag{5.2}$$

where g^{ij} is the inverse of g_{ij}. A similar reasoning is used to obtain the expression of the divergence of a vector field $\mathbf{X}$ on $\mathbb{M}$

$$\text{div}\mathbf{X} = \frac{1}{\sqrt{g}} \partial_i (\mathbf{X}^i \sqrt{g}), \tag{5.3}$$

where g is the determinant of g^{ij}. Finally, combining these two operators we obtain the LB operator, which is the second order differential operator defined on scalar fields on $\mathbb{M}$ by

$$\Delta I = -\frac{1}{\sqrt{g}} \partial_j (\sqrt{g} g^{ij} \partial_i I) \tag{5.4}$$

5.2.2 Computing a Generic Metric on the Sphere

As explained in Sect. 1.2 omnidirectional images are formed in two steps. The first one projects a 3D point to the unitary sphere. Then this point is projected from the unitary sphere to the image plane through a variable projection point, which is determined by the geometry of the mirror (parameter ξ). If the system is calibrated (Puig et al. 2011), it is also possible like in any conventional camera, to map the catadioptric image to the unitary sphere.

In Bogdanova et al. (2007), Arican and Frossard (2010), the mapping from para-catadioptric images to the sphere is used for the computation of the differential operators explained before. This allows to process the spherical image directly using the image coordinates. Here, we extend that approach to all central catadioptric systems.

Consider the unitary sphere $\mathbb{S}^2$ (Fig. 5.1a). A point on $\mathbb{S}^2$ can be represented in cartesian and polar coordinates as

$$(X, Y, Z) = (\sin\theta\cos\varphi, \sin\theta\sin\varphi, \cos\theta) \tag{5.5}$$

The Euclidean element in cartesian and polar coordinates is defined as

$$dl^2 = dX^2 + dY^2 + dZ^2 = d\theta^2 + \sin^2\theta d\varphi^2 \tag{5.6}$$

Under the sphere camera model, a point on the sphere (θ, φ), is mapped to a point in polar coordinates (R, φ) in the image plane. The θ angle depends on the central catadioptric system we are dealing with, while φ remain the same. For example, a conventional perspective system is described with $\theta = \arctan(R)$ and a paracatadioptric system with $\theta = 2\arctan(\frac{R}{2})$. In the general case (see Fig. 5.1d) we have

$$\theta = \arctan\left(\frac{R\left(1 + \xi + \sqrt{(1+\xi)^2 - R^2(\xi^2 - 1)}\right)}{1 + \xi - R^2\xi + \sqrt{(1+\xi)^2 - R^2(\xi^2 - 1)}}\right) \tag{5.7}$$

In terms of these new coordinates the metric becomes

$$dl^2 = \frac{\left(\xi + \xi^2 + \sqrt{(1+\xi)^2 - R^2(\xi^2 - 1)}\right)^2\left(R^2 d\varphi^2 + \frac{(1+\xi)dR^2}{1 - R^2(\xi - 1) + \xi}\right)}{(R^2 + (1+\xi)^2)^2} \tag{5.8}$$

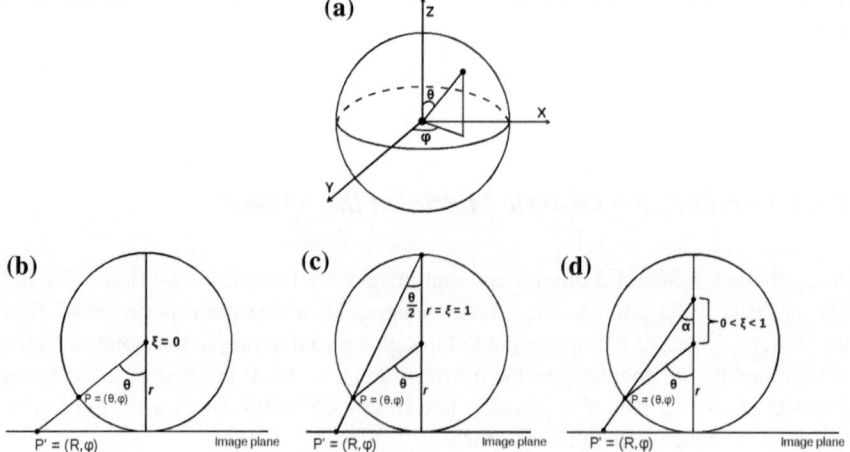

Fig. 5.1 Simplified sphere camera model, showing the different radial projection functions. **a** Spherical coordinates. **b** Perspective projection. **c** Paracatadioptric projection. **d** Hypercatadioptric projection

Let $(x, y) \in \mathbb{R}^2$ on the sensor plane define cartesian coordinates, where $R^2 = x^2 + y^2$ and $\varphi = \arctan(\frac{y}{x})$

$$dl^2 = \lambda \left(4(ydx - xdy)^2 - \frac{4(1 + \xi)(xdx + ydy)^2}{(x^2 + y^2)(\xi - 1) - \xi - 1} \right) \tag{5.9}$$

where

$$\lambda = \frac{\left(\xi + \xi^2 + \sqrt{(1 + \xi)^2 - (x^2 + y^2)(\xi^2 - 1)} \right)^2}{4(x^2 + y^2)(x^2 + y^2 + (1 + \xi)^2)^2}$$

To verify the correctness of this approach we substitute ξ to the values for which we already know the metric. For $\xi = 1$ the result is the same as that one presented in Arican and Frossard (2010). In the case of $\xi = 0$ that corresponds to the perspective case we expect a cartesian metric but what we get is the following

$$dl^2 = \frac{(1 + y^2)dx^2 - 2xydxdy + (1 + x^2)dy^2}{(1 + x^2 + y^2)^2} \tag{5.10}$$

This is explained since there is no map from the sphere to the plane both being conformal and area-preserving. However, from the generic metric we have already computed, we are able to compute the LB operator. To deal with the perspective case we only need to use the classical Laplacian.

5.2.3 Computing the Generic Metric on the Normalized Plane

The previous approach computes the metric on the sphere and then it is projected to the image plane through the angle θ which encodes the radial projection function. We have observed that using the sphere camera model we are able to compute the metric directly on the image plane and at the same time take into account the geometry of the mirror, which is given by the mirror parameter ξ.

Let's observe the perspective projection in cartesian and spherical coordinates

$$\left(\frac{X}{Z}, \frac{Y}{Z} \right) = \left(\tan \theta \cos \varphi, \tan \theta \sin \varphi \right) = (x, y) \tag{5.11}$$

The radial component is $R = \sqrt{x^2 + y^2} = \tan(\theta)$ and the Euclidean element in this case just takes into account the (x, y) coordinates but also the two spherical coordinates (θ, φ)

$$ds^2 = d(\tan \theta \cos \varphi)^2 + d(\tan \theta \sin \varphi)^2 = dx^2 + dy^2 \tag{5.12}$$

The same approach applied to the paracatadioptric case $\xi = 1$ gives

$$\left(\frac{X}{1-Z}, \frac{Y}{1-Z}\right) = \left(\frac{\sin\theta\cos\varphi}{1-\cos\theta}, \frac{\cos\theta\cos\varphi}{1-\cos\theta}\right) = (x, y) \tag{5.13}$$

and its corresponding Euclidean element with $\theta = 2\arctan\left(\frac{\sqrt{x^2+y^2}}{2}\right)$ and $\varphi = \arctan\left(\frac{y}{x}\right)$ becomes

$$dl^2 = \frac{16(dx^2 + dy^2)}{(x^2 + y^2 - 4)^2\left(\sqrt{\frac{(x^2+y^2+4)^2}{(x^2+y^2-4)^2}} - 1\right)^2} \tag{5.14}$$

The generic metric equation for any catadioptric system is described by the following Euclidean element on the normalized plane

$$ds^2 = \frac{(\xi\cos\theta - 1)^2 d\theta^2 + (\xi - \cos\theta)^2 \sin^2\theta d\varphi^2}{(\xi - \cos\theta)^4} \tag{5.15}$$

which allows to compute the Euclidean element on the normalized plane with the substitution of θ and φ in terms of x and y.

5.2.4 Generic Laplace–Beltrami Operator

From (5.9) we can compute the generic metric in matrix form g_{ij} and its corresponding inverse matrix g^{ij}

$$g^{ij} = \gamma_2\begin{pmatrix} -x^2(\xi - 1) + \xi + 1 & xy(\xi - 1) \\ xy(\xi - 1) & -y^2(\xi - 1) + \xi + 1 \end{pmatrix} \tag{5.16}$$

with

$$\gamma_2 = \frac{\left(x^2 + y^2 + (1 + \xi)^2\right)^2}{(1 + \xi)\left(\xi + \xi^2 + \sqrt{1 - (x^2 + y^2)(\xi^2 - 1) + 2\xi + \xi^2}\right)^2}$$

The determinant of (5.16) is

$$\det(g^{ij}) = -\frac{\left(x^2 + (1 + \xi)^2 + y^2\right)^4\left((x^2 + y^2)(\xi - 1) - \xi - 1\right)}{(1 + \xi)\left(\xi + \xi^2 + \sqrt{1 - (x^2 + y^2)(\xi^2 - 1) + 2\xi + \xi^2}\right)^4} \tag{5.17}$$

With all these elements we are now able to compute the LB operator (5.4) which is represented by the differential operators

Algorithm 1: Smoothing catadioptric images using the heat diffusion equation

Input : $I(x, y), t, d_t, \xi, Ix, Iy, Ixy, Ixx, Iyy$
Output: $I(x, y, t)$

Initialize required variables
[cIx,cIy,cIxy,cIxx,cIyy]$\leftarrow$ CDCoeff $(I(x, y), \xi)$
$ntimes \leftarrow t/d_t$
$I(x, y, t) \leftarrow I(x, y)$
for $i \leftarrow 1$ **to** $ntimes$ **do**
    ImIx $\leftarrow$ cIx *Convolve $(I(x, y, t), Ix)$
    ImIy $\leftarrow$ cIy *Convolve $(I(x, y, t), Iy)$
    ImIxy $\leftarrow$ cIxy *Convolve $(I(x, y, t), Ixy)$
    ImIxx $\leftarrow$ cIxx *Convolve $(I(x, y, t), Ixx)$
    ImIyy $\leftarrow$ cIyy *Convolve $(I(x, y, t), Iyy)$
    LBO $\leftarrow$ ImIx + ImIy+ImIxy+ImIxx + ImIyy
    $I(x, y, t) \leftarrow I(x, y, t) + d_t$ * LBO
end

$$\Delta I = \lambda_1 \frac{\partial^2 I}{\partial x^2} + \lambda_2 \frac{\partial^2 I}{\partial y^2} + \lambda_3 \frac{\partial^2 I}{\partial xy} + \lambda_4 \frac{\partial I}{\partial x} + \lambda_5 \frac{\partial I}{\partial y} \tag{5.18}$$

$$\lambda_1 = \frac{(x^2(\xi^2 - 1) - 1)(1 + x^2 + y^2)^2}{-1 - \xi^2 + (\xi^2 - 1)(x^2 + y^2) - 2\xi\sqrt{1 - (\xi^2 - 1)(x^2 + y^2)}} \tag{5.19}$$

$$\lambda_2 = \frac{(y^2(\xi^2 - 1) - 1)(1 + x^2 + y^2)^2}{-1 - \xi^2 + (\xi^2 - 1)(x^2 + y^2) - 2\xi\sqrt{1 - (\xi^2 - 1)(x^2 + y^2)}} \tag{5.20}$$

$$\lambda_3 = \frac{2xy(\xi^2 - 1)(1 + x^2 + y^2)^2}{-1 - \xi^2 + (\xi^2 - 1)(x^2 + y^2) - 2\xi\sqrt{1 - (\xi^2 - 1)(x^2 + y^2)}} \tag{5.21}$$

$$\lambda_4 = \frac{4x(1 + x^2 + y^2)\left(2 - \xi^2 - 2(\xi^2 - 1)(x^2 + y^2) - \xi\sqrt{1 - (\xi^2 - 1)(x^2 + y^2)}\right)}{-1 - \xi^2 + (\xi^2 - 1)(x^2 + y^2) - 2\xi\sqrt{1 - (\xi^2 - 1)(x^2 + y^2)}} \tag{5.22}$$

$$\lambda_5 = \frac{4y(1 + x^2 + y^2)\left(2 - \xi^2 - 2(\xi^2 - 1)(x^2 + y^2) - \xi\sqrt{1 - (\xi^2 - 1)(x^2 + y^2)}\right)}{-1 - \xi^2 + (\xi^2 - 1)(x^2 + y^2) - 2\xi\sqrt{1 - (\xi^2 - 1)(x^2 + y^2)}} \tag{5.23}$$

An analogous process has been performed for the computation of the metric in the normalized plane, but the representation of the equations is not suitable to show in these pages. The code used to generate the equations can be downloaded from my website.[1] From now on we use the metric computed on the sphere.

---

[1] http://webdiis.unizar.es/~lpuig/Mathematica_code/Mathematica_code_LBO.tar.gz

5.2.5 Smoothing Catadioptric Images Using the Generic Laplace–Beltrami Operator

We compute the smoothing of the catadioptric images using the heat diffusion equation. This equation is computed at successive time steps, $t_i = k^{2i}\sigma_o^2$ is defined in terms of the normalization and scale factors k and the base smoothing level σ_o.

The differentiation with respect to time in the heat diffusion equation is discretized with time intervals, d_t. To compute the discrete differential representations of the image $\frac{\partial I}{\partial x}$, $\frac{\partial I}{\partial y}$, $\frac{\partial^2 I}{\partial xy}$, $\frac{\partial^2 I}{\partial x^2}$ and $\frac{\partial^2 I}{\partial y^2}$ we convolve the catadioptric image with different kernels

$$
I_x = [-1\ 1],\ I_y = \begin{bmatrix} -1 \\ 1 \end{bmatrix},\ I_{xx} = [1\ -2\ 1],\ I_{yy} = \begin{bmatrix} 1 \\ -2 \\ 1 \end{bmatrix},\ I_{xy} = \begin{bmatrix} 0 & 1 & 0 \\ 1 & -4 & 1 \\ 0 & 1 & 0 \end{bmatrix}
$$

(5.24)

For a particular catadioptric system defined by ξ we compute the corresponding coefficients (5.19–5.23) which multiply its corresponding convolved image. Then we compute the LB operator. Finally, smoothing is performed by updating $I(x, y, t)$ with the differences computed at previous time steps.

5.3 Experiments

In this section, we perform several experiments using synthetic and real images to evaluate the scale-space for different catadioptric systems. The synthetic catadioptric images are generated using the raytracing software POV-ray[2] and correspond to images from an office scene.[3] Two hypercatadioptric systems with mirror parameters of $\xi = 0.9662$[4] (m1) and $\xi = 0.7054$[5] (m2) are considered. We also consider one paracatadioptric system $\xi = 1$ with radius $r = 2.5$ cm. The real images are acquired using the hypercatadioptric system designed by Neovision (m1).

5.3.1 Smoothing of Synthetic Images

We follow the Algorithm 1 with $d_t = 0.01$. All the coefficients $\lambda_i, i = 1 \ldots 5$ are computed once since they only depend on the geometry of the image and not on

---

[2] http://www.povray.org

[3] http://www.ignorancia.org

[4] http://www.neovision.cz

[5] http://www.accowle.com

the gray values. Figure 5.2 shows smoothed images with different mirrors at the same scale factor $t = 3$. They also shows the corresponding generic LB operator. We observe that the geometry of each mirror is taken into account when the smoothing is performed. The LB operator of m1 and the one corresponding to the paracatadioptric system is similar since the mirror parameter is close to 1. In the LB operators of mirrors m1 and m2, we observe how the intensities and thickness of the edges vary with respect to the distance to the image center. The differences between these LB operators are explained since the geometries of these mirrors are different.

In order to verify the validity of our approach we perform an experiment where we compare the generic LB operator computed using our approach to the cartesian Laplace operator. To obtain this operator we smooth the omnidirectional image to the

Fig. 5.2 Smoothed catadioptric images with $t = 3$. The first column corresponds to mirror m1, the second one to mirror m2, and the third one to the paracatadioptric system. The first row represents the original images, the second represents the smoothed images, and the last one the corresponding LB operators

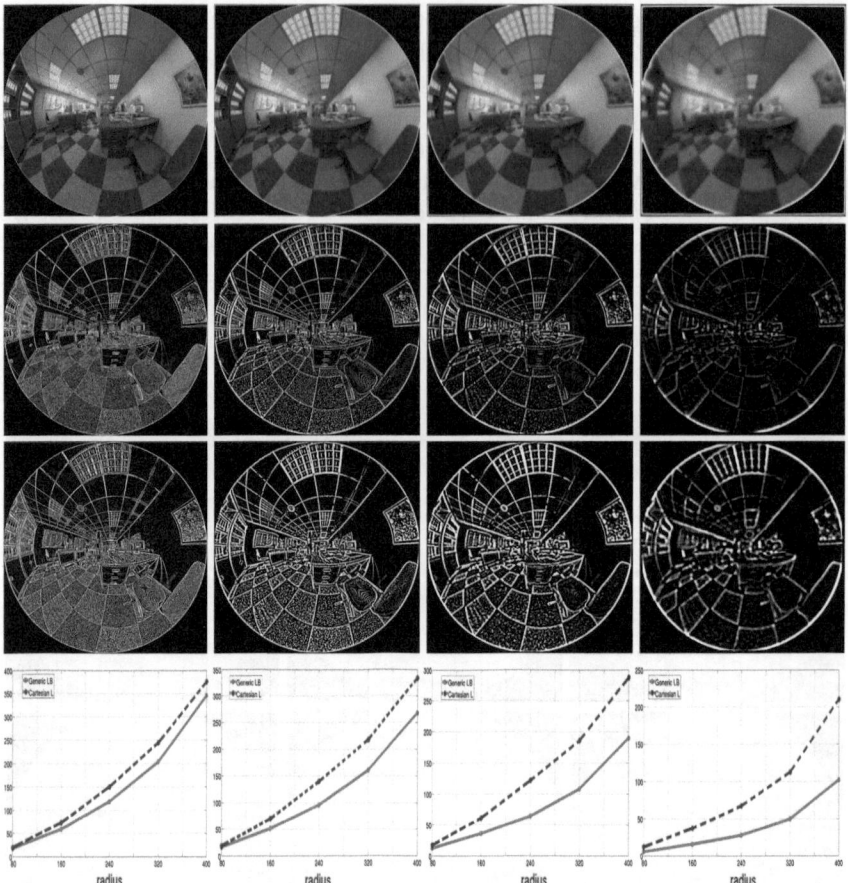

Fig. 5.3 Comparison between generic LB operator and cartesian Laplacian. The first row shows the smoothed images for scales $t = \{1, 2, 3, 5\}$ using our approach. The second row shows the corresponding generic LB operators. The third row, the cartesian Laplacians. The last row presents the sum of the values (scaled for visualization purposes) of the generic LB and the cartesian Laplacian as a function of the radius of a circle with origin at the image center

same scale t, using the corresponding Gaussian kernel. Then, we compute the Laplacian of this image using the cartesian operator I_{xy}. We select the hypercatadioptric (m1) image for which we compute scales $t = \{1, 2, 3, 5\}$ and their corresponding generic LB and cartesian Laplacian. In Fig. 5.3 we show the results. We observe that the generic LB operator considers a difference between the pixels close to the center and those close to the periphery, while the cartesian Laplacian has the same effect on all pixels, without taking into account their location with respect to the center.

To quantify the last statement we sum the values of the generic LB operator inside circles with different radii and origin at the image center. We perform the same procedure with the cartesian Laplacian. The last row of Fig. 5.3 shows the comparison

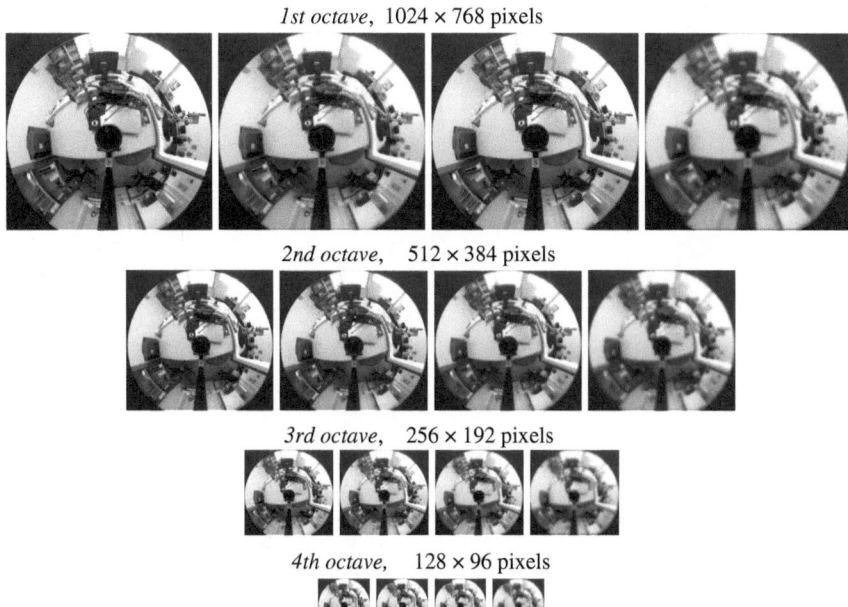

Fig. 5.4 Scale-Space of a real omnidirectional image, composed of four octaves and four scales per octave

of these sums for the different scales. We observe that the pixels in the periphery have smaller values in the generic LB operator than those on the normal Laplace operator. In the center where the resolution of the catadioptric images is bigger the values of the LB operator and the normal Laplace are similar.

5.3.2 Scale-Space on Real Catadioptric Images

In Fig. 5.4 we show the pyramid that compose the scale space of a hypercatadioptric image computed using our approach. We define the scale-space to be composed of four octaves and four scales per octave. The initial sigma is defined as $t = 0.8$ and the rest are computed as explained in Sect. 5.2.5. The value of the smoothing interval is defined as $k = 2^{1/3}$.

5.3.3 Repeatability Experiment

In this experiment we test the repeatability of the extrema points detected with the scale-space computed using our approach. The extrema points are obtained from the

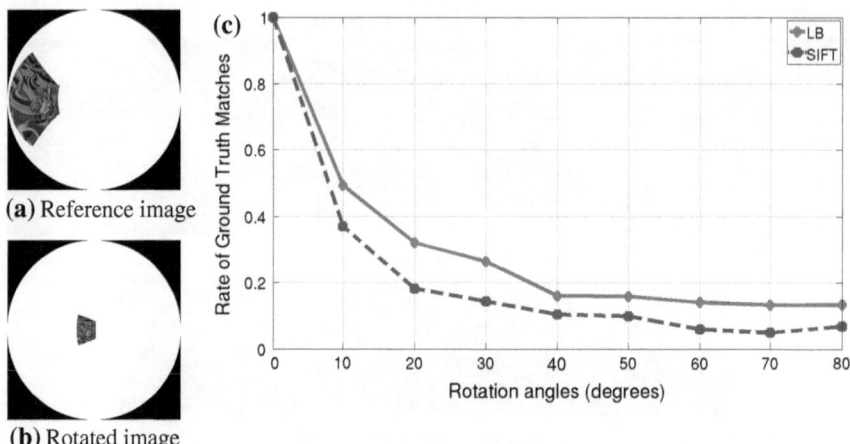

(a) Reference image

(b) Rotated image

Fig. 5.5 Repeatability experiment using synthetic hypercatadioptric images. **a** Reference image. **b** Image generated with the catadioptric system rotated 80° around the z-axis. **c** Percentage of correct matches through rotations of the catadioptric system around the z-axis. The x-axis represents the orientation of the catadioptric system in degrees with respect to the initial position

approximation to the scale-normalized Laplacian of Gaussian through differences of Gaussians, similar to Lowe (2004). We generate nine synthetic images with rotations from zero degrees to eighty degrees around the z-axis, with steps of ten degrees between each pair of images. The two extreme images are shown in Fig. 5.5. We observe a drastic distortion on the images, this is produced by the conic shape of the mirror.

Since the rotation of the catadioptric system is known, we can map the detected features in the reference image $\bar{\mathbf{x}}$ to the subsequent images and compute the distance between them. The matching criteria is the following

$$\|\hat{\mathbf{x}} - \bar{\mathbf{x}}\| \leq \delta_d \qquad (5.25)$$

where, $\hat{\mathbf{x}}$ is the mapped point. We use the Euclidean distance. The distance threshold δ_d has to be adapted to capture the matches with different scale levels, which is $\delta_d = \delta_0 \cdot t_{\bar{\mathbf{x}}}$ with δ_0 the distance threshold parameter.

We compare our approach to SIFT algorithm. In particular we use Vedaldi's implementation (Vedaldi 2007). In Fig. 5.5c we show the results of this experiment. We observe that the LB approach has a clear advantage over the classical SIFT approach, obtaining double repeatability than the scale-space used by SIFT in extreme camera rotations. The low rate shows the difficulties of matching catadioptric images.

5.4 Closure

We have presented a new way to compute the scale-space of omnidirectional images. We integrate the sphere camera model which considers all central catadioptric systems with the partial differential equation framework on manifolds to compute a generic version of the second order differential operator Laplace–Beltrami. This operator is used to perform the Gaussian smoothing on catadioptric images. We perform experiments using synthetic images generated with parameters coming from actual manufactured mirrors. We observe that LB operator considers correctly the geometry of the mirror, since the pixels at the periphery have a different weight than those at the center. This situation explains the natural nonhomogeneous resolution inherent to the central catadioptric systems. The near future work is to implement a complete scale-invariant feature detector also invariant to camera since the sphere camera model allows to consider all central projections.

Chapter 6
Orientation of a Hand-Held Catadioptric System in Man-Made Environments

Abstract In this chapter, we present a new approach to extract conics from catadioptric images, in particular those which represent the projection of 3D lines in the scene. Using the internal calibration and two image points, we are able to compute the catadioptric image lines analytically. The presence of parallel lines in man-made environments is exploited to compute the dominant vanishing points in the omnidirectional image. The vanishing points are extracted from the intersection of two conics in the catadioptric image, which represent parallel lines in the scene. This intersection is performed via the common self-polar triangle associated to this pair. From the information contained in the vanishing points, we obtain the orientation of a hand-held hypercatadioptric system. This approach is tested by performing vertical and full rectifications in real sequences of images.

6.1 Introduction

In robotics, when a catadioptric system is used, it is commonly observed that it has a vertical orientation. This is because most robotic platforms used are wheel-based. Under this configuration, planar-motion and/or 1D image geometry is assumed which reduces the degrees of freedom (DOF) of the problem (Guerrero et al. 2008). Moreover, in applications where line tracking or line matching is performed, this assumption is useful (Mezouar et al. 2004; Murillo et al. 2007). Besides that there exist robot platforms where the vertical assumption is not satisfied, and they require the development of new algorithms to interact with the environment. One of these algorithms can be a self-orientation system to be used for the stabilization of a biologically inspired humanoid robot platform (Miyauchi et al. 2007). One of the advantages of the nonvertical configuration is that both the horizontal and vertical vanishing points are present in the image and can be computed by the intersection of parallel lines. In man-made environments, we can observe sets of parallel and orthogonal lines and planes that can be exploited to compute the orientation of the

L. Puig and J. J. Guerrero, *Omnidirectional Vision Systems*, SpringerBriefs in Computer Science, DOI: 10.1007/978-1-4471-4947-7_6, © Luis Puig 2013

system (Kosecka and Zhang 2002). However, the extraction of lines in catadioptric images becomes extraction of conics. Five collinear image points are required to extract them in the uncalibrated case. However, two points are enough if we take advantage of the internal calibration of the catadioptric system. We call these lines, catadioptric image lines (CILs). Some works have been proposed to deal with this problem. In (Vasseur and Mouaddib 2004), the space of the equivalent sphere which is the unified domain of central catadioptric sensors combined with the Hough transform is used. In (Ying and Hu 2004b), they also use the Hough transform and two parameters on the Gaussian sphere to detect the image lines. The accuracy on the detection of these two approaches depends on the resolution of the Hough transform. The higher the accuracy, the more difficult to compute the CILs. In (Mei and Malis 2006), the randomized Hough transform is used to overcome the singularity present in (Vasseur and Mouaddib 2004; Ying and Hu 2004b) and to speed up the extraction of the conics. This scheme is compared in converge mapping to a RANSAC approach. In (Bazin et al. 2007), a scheme of split and merge is proposed to extract the CILs present in a connected component. These connected components, as in our case, are computed in two steps. The first step consists of detecting the edges using the Canny operator. The second step is a process of chaining which builds the connected components. In contrast to (Vasseur and Mouaddib 2004; Ying and Hu 2004b; Mei and Malis 2006), our approach does not use the Hough transform; instead, we compute the CIL directly from two image points present in a connected component. Then, a RANSAC approach is used to identify the points that belong to this conic. As opposed to (Bazin et al. 2007), we use an estimation of the geometric distance from a point to a conic instead of an algebraic distance. Notice that a connected component can contain more than one CIL and the process has to be repeated until all CILs are extracted.

Once we have extracted the lines in the catadioptric images (CILs), we need to compute the intersection of parallel CILs to extract the vanishing points. In this work—previously presented in (Puig et al. 2010)—we propose a modification to the computation of the common self-polar triangle (Barreto 2003) in order to compute the intersection between a pair of CILs. Instead of having four intersections points between two general conics, we have just two in the case of CILs. When this intersection corresponds to parallel CILs, these points are the vanishing points. We compute all the intersection between the CILs present in the image. Then with a voting approach, we robustly determine which ones are the vanishing points. The first vanishing point to compute is the vertical vanishing point (VVP) from which we are able to perform a rectification of the omnidirectional image. With this rectification, we obtain an omnidirectional image fitting the vertical assumption and the applications designed with this constraint can be used. Using an analogous process, we compute the horizontal vanishing point (HVP). From this HVP, we compute the last angle that gives the whole orientation of the catadioptric system.

6.2 Catadioptric Image Lines (CIL) Computing

In this section, we explain the method used to extract the CILs from two image points. As mentioned before in the case of uncalibrated systems, we require five points to describe a conic. If these points are not distributed in the whole conic, the estimation is not correctly computed. Another disadvantage of a 5-point approach is the number of parameters. When a robust technique is used, like RANSAC, this is quite important because the number of iterations required hardly increases with the number of parameters of the model. Our approach overcomes these problems since two points are enough. As we assume, the calibrated camera can describe the conics using only two parameters and the calibration parameters, which allows to extract the CIL from two points. We compute the points in the normalized plane $\mathbf{r} = (sx \; sy \; s)^\mathsf{T} = (x \; y \; 1)^\mathsf{T}$ using the inverse of matrix K

$$\mathbf{r} = \mathsf{K}^{-1}\mathbf{q}. \tag{6.1}$$

As previously presented, a 2D line $\mathbf{n}$ in the normalized plane, which corresponds to the projection of a 3D line under the sphere camera model, can be represented as $\mathbf{n} = (n_x, n_y, n_z)^\mathsf{T}$. The points $\mathbf{r}$ in the normalized plane lying on this line satisfy $\mathbf{r}^\mathsf{T}\Omega\mathbf{r} = 0$, where Ω is defined as

$$\Omega = \begin{pmatrix} n_x^2 (1 - \xi^2) - n_z^2\xi^2 & n_x n_y (1 - \xi^2) & n_x n_z \\ n_x n_y (1 - \xi^2) & n_y^2 (1 - \xi^2) - n_z^2\xi^2 & n_y n_z \\ n_x n_z & n_y n_z & n_z^2 \end{pmatrix} \tag{6.2}$$

Developing the relation $\mathbf{r}^\mathsf{T}\Omega\mathbf{r} = 0$ and after some algebraic manipulation we obtain

$$(1 - \xi^2)(n_x x + n_y y)^2 + 2n_z(n_x x + n_y y) + n_z^2(1 - \xi^2(x^2 + y^2)) = 0 \tag{6.3}$$

simplifying

$$(1 - \xi^2 r^2)\beta^2 + 2\beta + (1 - \xi^2) = 0 \tag{6.4}$$

where a change of variable to $\beta = \frac{n_z}{n_x x + n_y y}$ and $r^2 = x^2 + y^2$ is performed. We can compute β by solving the quadratic equation

$$\beta = -\frac{1}{1 - \xi^2 r^2} \pm \frac{\xi}{1 - \xi^2 r^2}\sqrt{1 + r^2 (1 - \xi^2)} \tag{6.5}$$

After solving this quadratic equation, we compute the normal $\mathbf{n}$. Consider two points in the normalized plane $\mathbf{x}_1 = (x_1, y_1, 1)^\mathsf{T}$ and $\mathbf{x}_2 = (x_2, y_2, 1)^\mathsf{T}$. From (6.5), we compute the corresponding β_1 and β_2. Notice that there exist two solutions for β

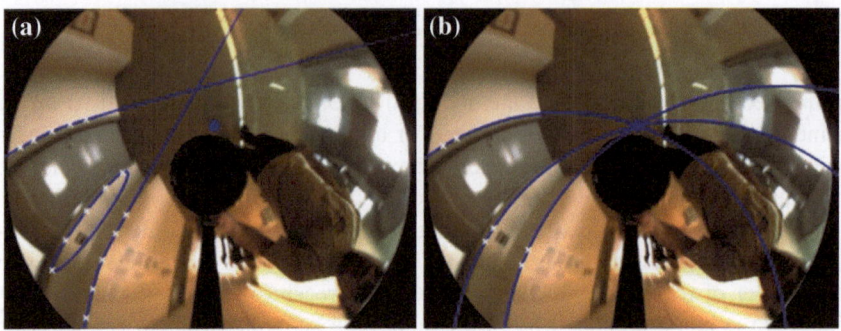

Fig. 6.1 Computing a CIL with (**a**) using the five point approach. (**b**) using our approach with only two close points. The central blue point corresponds to the vertical vanishing point

and just one has a physical meaning.[1] Using these parameters, we obtain the linear system

$$\begin{pmatrix} x_1 & y_1 & -\dfrac{1}{\beta_1} \\ x_2 & y_2 & -\dfrac{1}{\beta_2} \end{pmatrix} \begin{pmatrix} n_x \\ n_y \\ n_z \end{pmatrix} = \begin{pmatrix} 0 \\ 0 \end{pmatrix} \tag{6.6}$$

Since **n** is orthonormal $n_x^2 + n_y^2 + n_z^2 = 1$. Solving for n_x, n_y and n_z, we have

$$n_x = \frac{y_1/\beta_2 - y_2/\beta_1}{v}, \qquad n_y = \frac{x_2/\beta_1 - x_1/\beta_2}{v}, \qquad n_z = \frac{x_2 y_1 - x_1 y_2}{v}, \tag{6.7}$$

with $v = \sqrt{(x_2 y_1 - x_1 y_2)^2 + (y_1/\beta_2 - y_2/\beta_1)^2 + (x_2/\beta_1 - x_1/\beta_2)^2}$.

Notice that we have analytically computed the normal **n** that defines the projection plane of the 3D line. In Fig. 6.1 we show a comparison of the computing of a image line in the uncalibrated case using five points, and the calibrated case using our approach with only two points. In this figure, we can observe that our approach obtains a better estimation even with two very close points. We also observe that the distance of the conic to the vanishing point using our two-point approach is much better than the general 5-point approach.

.

6.2.1 Catadioptric Image Lines Extraction

Our line extraction proposal can be explained as follows. First we detect the edges using the Canny algorithm. Then the connected pixels are stored in components. For each component, we perform a RANSAC approach to detect the CILs present in this

---

[1] We have observed that the negative solution is the correct one.

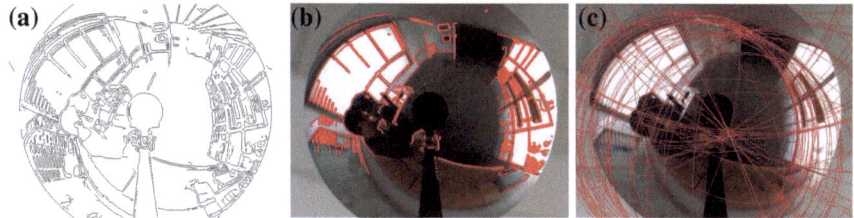

Fig. 6.2 Extraction of image lines (CILs). **a** Canny edge detector result, **b** connected components, and **c** CILs extracted

component. Two points from the connected component are chosen randomly and the corresponding CIL is computed. The distance from the rest of the points to this CIL is computed. The points with a distance smaller than some threshold vote for this CIL. The process stops when the number of points that has not voted for any conic and the number of points in the component are smaller than a threshold. In Fig. 6.2, we can observe the three main steps to extract the CILs.

6.2.2 Distance from a Point to a Conic

In order to know if a point $\mathbf{x}$ lies on a conic $\mathbf{C}$, we need to compute the distance from a point to a conic. Two distances are commonly used to this purpose. The algebraic distance defined by (6.8) which just gives an scalar value and the geometric distance, which gives the distance from this point to the closest point on the conic. This distance is more difficult to calculate and its computing is time consuming. We propose an estimation to this distance replacing the point-to-conic distance by a point-to-point distance (see Fig. 6.3). Our proposal is based on the gradient of the algebraic distance from a point $\mathbf{x}_c$ to a conic represented as a six-vector $\mathbf{C} = (c_1, c_2, c_3, c_4, c_5, c_6)$

$$d_{\text{alg}} = c_1 x^2 + c_2 xy + c_3 y^2 + c_4 x + c_5 y + c_6. \tag{6.8}$$

We define the perpendicular line to a point that lies on the conic $\mathbf{C}$ as

$$\ell_\perp = \mathbf{x}_c + \lambda \tilde{\mathbf{n}}(\mathbf{x}_c), \quad \text{where} \quad \tilde{\mathbf{n}}(\mathbf{x}_c) = \frac{\nabla d_{\text{alg}}}{\|\nabla d_{\text{alg}}\|} \tag{6.9}$$

The normal vector $\tilde{\mathbf{n}}$ is computed from the gradient of the algebraic distance.

$$\nabla d_{\text{alg}} = \begin{pmatrix} \dfrac{\partial f}{\partial x} \\ \dfrac{\partial f}{\partial y} \end{pmatrix} = \begin{pmatrix} 2c_1 x + c_2 y + c_4 \\ c_2 x + 2c_3 y + c_5 \end{pmatrix} \tag{6.10}$$

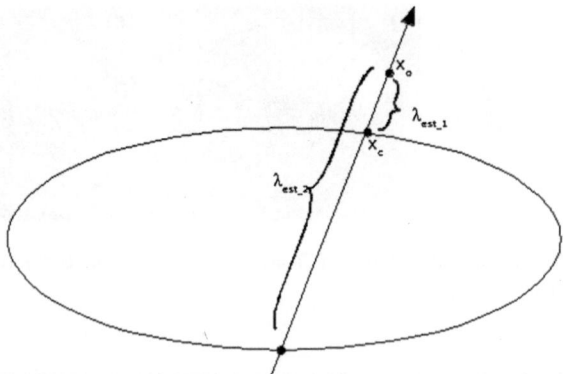

Fig. 6.3 Approximation to the distance from a point to a conic

When a point does not lie on the conic $\mathbf{x}_o$, we can compute an estimation to its corresponding perpendicular line using the property that $\tilde{\mathbf{n}}(\mathbf{x}_c) = \tilde{\mathbf{n}}(\mathbf{x}_o + \Delta \mathbf{x}) \approx \tilde{\mathbf{n}}(\mathbf{x}_o)$

$$\ell_{\text{est}} = \mathbf{x}_o + \lambda_{\text{est}}\tilde{\mathbf{n}}(\mathbf{x}_o) = \begin{pmatrix} x_o + \lambda_{\text{est}}\tilde{n}_x(\mathbf{x}_o) \\ y_o + \lambda_{\text{est}}\tilde{n}_y(\mathbf{x}_o) \end{pmatrix} \tag{6.11}$$

To compute λ_{est}, we substitute x by $x_o + \lambda_{\text{est}}\tilde{n}_x(\mathbf{x}_o)$ and y by $y_o + \lambda_{\text{est}}\tilde{n}_y(\mathbf{x}_o)$ in (6.8), giving a quadratic equation

$$\lambda_{\text{est}}^2 \underbrace{(c_1\tilde{n}_x^2 + c_2\tilde{n}_x\tilde{n}_y + c_3\tilde{n}_y^2)}_{a} + \lambda_{\text{est}} \underbrace{(2c_1\tilde{n}_x + 2c_3\tilde{n}_y + c_2(x_0\tilde{n}_y + y_0\tilde{n}_x))}_{b}$$
$$+ \underbrace{c_1x_0^2 + c_2x_0y_0 + c_3y_0^2 + c_4x_0 + c_5y_0 + c_6}_{c} = 0 \tag{6.12}$$

We observe that λ_{est} gives the two distances that intersect the conic; so we choose the closest to $\mathbf{x}_o$ as the distance from that point to the conic $d = \|\mathbf{x}_o - \mathbf{x}_c\| = \lambda_{\text{est}}$.

6.3 Intersection of Two CILs

In a general configuration, two conics intersect in four points. The intersection of these points defines three distinct pair of lines. The intersection of these lines represents the vertices of the self-polar triangle common to a pair of conics (Barreto, 2003). We study the particular case where two CILs intersect, which is a degenerate configuration with just two intersection points. As we observe in Fig. 6.4, there exists a line r that intersects these two points and the origin of the normalized plane. Our goal is to compute this line and from it to extract the two intersections of the conics that correspond to points $\mathbf{P}^+$ and $\mathbf{P}^-$.

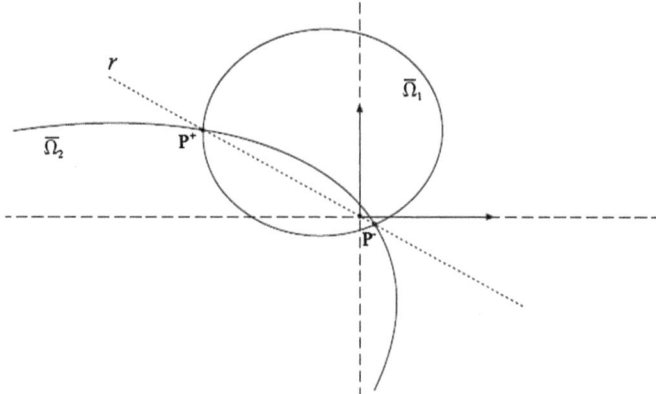

Fig. 6.4 Intersection of two CILs in the normalized plane

Let $\mathbf{n}_1 = (n_{x_1}, n_{y_1}, n_{z_1})^{\mathsf{T}}$, and $\mathbf{n}_2 = (n_{x_2}, n_{y_2}, n_{z_2})^{\mathsf{T}}$ the two normal vectors represent the projection of two lines in the scene and Ω_1 and Ω_2 the two conics represent the image lines in the normalized plane. The vertices of the self-polar triangle associated to the pencil $\Omega(\lambda) = \Omega_1 + \lambda\Omega_2$ satisfy

$$\det(\Omega_1 + \lambda\Omega_2) = 0. \tag{6.13}$$

If we develop this constraint, we obtain a third order polynomial where just one of the solutions is real and it corresponds to $\lambda_1 = -n_{z_1}^2/n_{z_2}^2$. So, the null space of $\Omega(\lambda_1) = \Omega_1 + \lambda_1\Omega_2$ is the line r, expressed in a parametric way as

$$r = \mu \cdot \mathbf{v} = \mu \begin{pmatrix} v_x \\ v_y \end{pmatrix} = \mu \begin{pmatrix} n_{z_2}^2 n_{y_1} n_{z_1} - n_{z_1}^2 n_{y_2} n_{z_2} \\ n_{z_1}^2 n_{x_2} n_{z_2} - n_{z_2}^2 n_{x_1} n_{z_1} \end{pmatrix}. \tag{6.14}$$

The intersection of this line to both Ω_1 and Ω_2 gives the two points $\mathbf{P}^+$ and $\mathbf{P}^-$. To obtain them, we solve for μ in (6.15) and substitute in (6.14).

$$\mu^2(c_1 v_x^2 + c_2 v_x v_y + c_3 v_y^2) + \mu(c_4 v_x + c_5 v_y) + c_6 = 0. \tag{6.15}$$

6.4 Vanishing Points

The vanishing points indicate the intersection of image lines corresponding to parallel lines in the scene. In vertical aligned catadioptric systems, vertical lines are radial lines in the image representation. Their intersection point, the vertical vanishing point (VVP), is located at the image center. When the camera is not vertically aligned, the radial lines become conic curves. In this case, the VVP moves from the image center and its new location contains important information about the orientation of the camera with respect to the scene.

We use a classic algorithm to detect the VVP. Let m be the number of putative vertical CILs detected in the omnidirectional image, and let $\mathbf{n}_i$ be their corresponding representation in the normalized plane. For every pair of CILs (there is a total of $m(m-1)/2$ pairs), we compute their intersection as explained above. Then for each line $\mathbf{n}_i$, we compute the distance to these points. If the line is parallel to that pair of CILs, the distance is smaller than a threshold and then that line is voted as VVP possible. The most voted point is considered the VVP.

6.5 Image Rectification

Here we explain the relation between the VVP computed in the normalized plane and the orientation of the catadioptric system. Writing the VVP in polar coordinates $\mathbf{x}_{vp} = (\rho_{vp}, \theta_{vp})^{\mathsf{T}}$ (see Fig. 6.5(d)) we observe that there exists a relation between the angle θ_{vp} and the angle ϕ representing the rotation of the catadioptric system around the z-axis (6.16). The negative angle is produced by the mirror effect which inverts the catadioptric image.

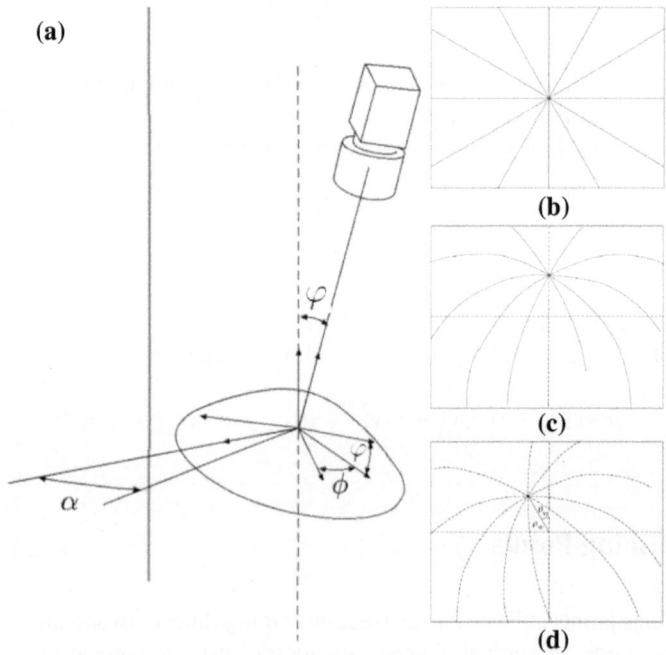

Fig. 6.5 **a** Configuration of the catadioptric system in a hand-held situation. **b** The VVP is in the center of the image. **c** The VVP moves in the vertical axis when the camera rotates around the x-axis. **d** The VVP rotates around the image center when the camera rotates around the z-axis

$$\phi = -\theta_{vp} \tag{6.16}$$

We observed that the component ρ_{vp} is intrinsically related to the rotation angle φ and the mirror parameter ξ of the catadioptric system. Since angles φ and ϕ are independent, we consider the case where $\phi = 0$ (see Fig. 6.5(c)). Using (6.14) and (6.15) with a pair of parallel CILs in polar coordinates, we compute the following relationship

$$\rho_{vp} = -\frac{\sin \varphi}{\cos \varphi \pm \xi}. \tag{6.17}$$

A lookup table can be built to speed up the computing of the vertical orientation. An analogous process is performed to detect the horizontal vanishing point. With the information provided by this point, we are able to compute the full orientation of the catadioptric system.

6.6 Experiments

In this section, we present some experiments rectifying real images. We acquire an image sequence with a calibrated hand-held hypercatadioptric system.[2] The calibration was performed using (Mei and Rives 2007). The process to extract the vanishing points and to perform the rectification can be summarized as follows: (i) the edges are detected by the Canny operator; (ii) the process to construct the connected components is performed; (iii) a RANSAC approach is performed for each connected component to extract all CILs present on it; (iv) all CIL intersections are computed and the vanishing points are estimated; (v) the vertical correction is performed using the orientation information obtained from the VVP; (vi) The full rectification is performed using the orientation information contained in the HVP.

In Fig. 6.6 we show an example of full rectification using our approach for a single frame of the image sequence. We observe how the VVP and the HVP are computed. Once the rectification angles are computed, we align the image to the reference system given by the scene itself. This allows to see how the only movement present in the sequence is translation (Fig. 6.7).

6.7 Closure

In this chapter, we presented a new way to extract lines in omnidirectional images generated by a calibrated catadioptric system. We only use two points lying on the CIL and an approximation to the geometric distance from a point to a conic. We also showed how to compute the intersection of two image lines based on the common

---

[2] http://www.neovision.cz/

Fig. 6.6 Example of full image rectification. **a** Frame 1 of the image sequence. **b** Conic extraction using our approach. **c** Putative vertical and horizontal vanishing points. *Yellow* circles represent putative vertical vanishing points. The *blue* ones represent the putative horizontal vanishing points and the *green* ones are the intersections points that cannot be consider either vertical or horizontal vanishing points. The *white* square is the estimated HVP and the *black* one is the VVP. **d** Full-rectified image. The vertical CILs are shown in white and the horizontal ones in *red*

Frame 107   Frame 242

Fig. 6.7 Panoramic representation of two full-rectified frames. *Vertical lines* are shown in white and horizontal ones in *red*. The horizontal vanishing point is aligned to the image center

self-polar triangle. From the intersection of CILs, we compute the vertical and horizontal vanishing points, which contain the orientation information of the catadioptric system with respect to the scene. To show the effectiveness of this approach, we perform experiments with real images. We compute the orientation of a hand-held hyper-catadioptric system through a video sequence.

Chapter 7
Conclusions

In this book we have studied omnidirectional cameras, in particular central catadioptric systems. We analyzed from the very basic step of calibration, where we propose a new method that is able to calibrate all central catadioptric systems including the pin-hole camera. In practice this method showed that it could go further, being able to calibrate noncentral catadioptric systems, such as the popular wide-angle fish-eye cameras. We also performed a deep analysis of the available calibration methods for central catadioptric systems, considering also those used to calibrate fish-eye cameras. The analysis is theoretical and practical. First, we present a classification of such methods and give the most relevant information about them. For the practical part we select those methods available on-line as OpenSource, since in the computer vision and robotics communities, these methods can save time and effort when the goal is beyond the calibration itself. The new calibration method and the analysis of calibration methods were presented in Chap. 2 and Chap. 3, respectively.

In Chap. 4 we studied in great detail the two-view relations of central catadioptric systems, in particular the combination of conventional cameras and catadioptric systems. We select this combination since a single catadioptric view contains a more complete description of the scene, and the perspective image gives a more detailed description of the particular area or object we are interested in. We perform a deep analysis on the behavior of three approaches to compute the hybrid fundamental matrix and the hybrid planar homography. From the simplest, specific for a particular mirror shape, to the generic ones, which consider all central catadioptric systems. We show how to extract relevant information from the fundamental matrix, more specific the extraction of the epipoles from the corresponding hybrid fundamental matrix. We also show how to impose the rank 2 property to this matrix, which is required to obtain the epipoles. We observed that the generic models are more accurate under ideal conditions, but at the same time they are more sensitive to perturbations (image noise) than that the simpler models. With the more reliable and simplest models we successfully match perspective images with catadioptric views robustly.

We also studied in this book the scale-space for catadioptric systems. The scale-space computation is the most important step on the extraction of scale-invariant features. We presented a method which is able to compute this scale-space for all central catadioptric systems. We propose to combine the sphere camera model and the partial differential equations framework on manifolds. With this combination we compute the Laplace–Beltrami (LB) operator related to any parametric surface, in this case, the manifold which represents the mirror of the catadioptric system. The LB operator is used to smooth the catadioptric images in order to construct the scale-space. Once the scale-space is obtained the approximation to the Laplacian of Gaussians (LoG) through the Difference of Gaussians (DoG) is performed and the extrema points are detected. This approach and some results with synthetic and real images are presented in Chap. 5.

In Chap. 6 we compute the orientation of a hypercatadioptric system in a man-made environment. We take advantage of the presence of straight line elements in such environments. These elements are aligned to the dominant directions and their projections give the orientation of the system with respect to that reference system. In this chapter, we propose different solutions to the different steps comprised computing the orientation of the system. In first place, the extraction of the projections of straight lines in the scene from the catadioptric image becomes more complex, since these projections are no longer lines but conics. We propose a new strategy to extract catadioptric image lines. This approach requires the calibration of the system and only 2 points lying on the catadioptric image lines we want to extract. To decide if a point belongs or not to a certain conic we propose a point-to-point distance that approximates the distance from a point to a conic. With the lines extracted the next step is to compute the intersection between projections of parallel lines. In this order, we propose the use of the common self-polar triangle adapted for catadioptric image lines, which instead of having four intersections as in the general case, it has only two and its computation is simplified to a solution of a quadratic equation. Finally, a classic voting scheme is used to identify the intersections of parallel lines with more supporting lines, which represent the vertical and the horizontal vanishing point. From the computed vanishing points, the extraction of the orientation of the catadioptric system with respect to the absolute reference system given by the dominant directions is straightforward.

As we can observe from this book, omnidirectional images are facing the same problems as the conventional ones with a more complex geometry. The development of omnidirectional vision is at the beginning if compared to the development of the conventional one. Because of this situation all areas where omnidirectional sensors could be used must be explored. In particular we would like to highlight the following:

- In the case of catadioptric systems where the information of the mirror is required to take into account its geometry for further applications. This situation opens an opportunity where automatic methods to extract this information from the environment should be developed, instead of obtaining the information from an off-line calibration process.
- The development of the omnidirectional vision should consider models that can deal with the majority of projection systems. Facing at the same time the problems present in the conventional images and the more complex omnidirectional ones.

- From conventional vision we observe that the first goal is to find reliable and robust methods to solve the problems. The second goal is to speed up this methods or to design faster ones based on the previously developed ones. This should be the way to lead the development of the omnidirectional vision.

References

Abdel-Aziz, Y. I., & Karara, H. M. (1971). Direct linear transformation from comparator coordinates into object space coordinates in close-range photogrammetry. In *Symposium on Close-Range Photogrammetry* (pp.1–18).

Aliaga, D. (2001). Accurate catadioptric calibration for real-time pose estimation in room-size environments. In *Proceedings of the 8th IEEE International Conference on Computer Vision, ICCV 2001* (Vol. 1, pp. 127–134).

Arican, Z., & Frossard, P. (2010). OmniSIFT: Scale invariant features in omnidirectional images. In *Proceedings of the International Conference on Image Processing (ICIP)*. Hong Kong, China: IEEE.

Baker, S., & Nayar, S. (1999). A theory of single-viewpoint catadioptric image formation. *International Journal of Compututer Vision, 35*(2), 175–196.

Barreto, J. (2003). *General central projection systems: Modeling, calibration and visual servoing*. PhD thesis, University of Coimbra, Portugal.

Barreto, J. (2006). A unifying geometric representation for central projection systems. *Computer Vision and Image Understanding, 103*(3), 208–217.

Barreto, J., & Araujo, H. (2001). Issues on the geometry of central catadioptric image formation. *Proceedings of the Conference on Computer Vision and Pattern Recognition, CVPR 2001* (Vol. 2, pp. 422–427).

Barreto, J., & Araujo, H. (2002). Geometric properties of central catadioptric line images. *Proceedings of the 7th European Conference on Computer Vision-Part IV, ECCV '02* (pp. 237–251). London, UK.

Barreto, J., & Araujo, H. (2005). Geometric properties of central catadioptric line images and their application in calibration. *IEEE Transactions on Pattern Analysis and Machine Intelligence, 27*(8), 1327–1333.

Barreto, J., & Daniilidis, K. (2006). Epipolar geometry of central projection systems using veronese maps. In *IEEE Computer Society Conference on Computer Vision and Pattern Recognition, 2006* (Vol. 1, pp. 1258–1265).

Bartoli, A., & Sturm, P. (2004). Non-linear estimation of the fundamental matrix with minimal parameters. *IEEE Transactions on Pattern Analysis and Machine Intelligence, 26*(4), 426–432.

Bastanlar, Y., Puig, L., Sturm, P., Guerrero, J. J., & Barreto, J. (2008). Dlt-like calibration of central catadioptric cameras. In *Workshop on Omnidirectional Vision, Camera Networks and Non-Classical Cameras*. Marseille, France.

Bazin, J. C., Demonceaux, C., & Vasseur, P. (2007). Fast central catadioptric line extraction. In *IbPRIA '07: Proceedings of the 3rd Iberian conference on Pattern Recognition and Image Analysis* (Part II, pp. 25–32).

Bertalmío, M., Cheng, L.-T., Osher, S., & Sapiro, G. (2001). Variational problems and partial differential equations on implicit surfaces. *Journal of Computational Physics, 174*(2), 759–780.

Bogdanova, I., Bresson, X., Thiran, J.-P., & Vandergheynst, P. (2007). Scale space analysis and active contours for omnidirectional images. *IEEE Transactions on Image Processing, 16*(7), 1888–1901.

Buchanan, T. (1988). The twisted cubic and camera calibration. *Computer Vision, Graphics and Image Processing, 42*(1), 130–132.

Bulow, T. (2004). Spherical diffusion for 3d surface smoothing. *IEEE Transactions on Pattern Analysis and Machine Intelligence, 26*(12), 1650–1654.

Caglioti, V., Taddei, P., Boracchi, G., Gasparini, S., & Giusti, A. (2007). Single-image calibration of off-axis catadioptric cameras using lines. *Proceedings of the 11th IEEE International Conference on Computer Vision. ICCV 2007* (pp. 1–6).

Cauchois, C., Brassart, E., Drocourt, C., & Vasseur, P. (1999). Calibration of the omnidirectional vision sensor: Syclop. In *Proceedings of the IEEE International Conference on Robotics and Automation, 1999* (Vol. 2, pp. 1287–1292).

Chen, D., & Yang, J. (2005). Image registration with uncalibrated cameras in hybrid vision systems. In *WACV/MOTION* (pp. 427–432).

Chen, X., Yang, J., & Waibel, A. (2003). Calibration of a hybrid camera network. In *Proceedings of Ninth IEEE International Conference on Computer Vision* (Vol. 1, pp. 150–155).

Claus, D., & Fitzgibbon, A. W. (2005). A rational function lens distortion model for general cameras. In *Proceedings of the IEEE Conference on Computer Vision and Pattern Recognition* (pp. 213–219).

Courbon, J., Mezouar, Y., Eck, L., & Martinet, P. (2007). A generic fisheye camera model for robotic applications. In *IEEE/RSJ International Conference on Intelligent Robots and Systems, 2007. IROS 2007* (pp. 1683–1688).

Cruz, J., Bogdanova, I., Paquier, B., Bierlaire, M., & Thiran, J.-P. (2009). Scale invariant feature transform on the sphere: Theory and applications. Technical Report, EPFL.

Deng, X.-M., Wu, F.-C., & Wu, Y.-H. (2007). An easy calibration method for central catadioptric cameras. *Acta Automatica Sinica, 33*(8), 801–808.

Espuny, F., & Burgos Gil, J. (2011). Generic self-calibration of central cameras from two rotational flows. *International Journal of Computer Vision, 91*, 131–145.

Faugeras, O. (1993). *Three-dimensional computer vision (artificial intelligence)*. London: The MIT Press.

Frank, O., Katz, R., Tisse, C., & Durrant Whyte, H. (2007). Camera calibration for miniature, low-cost, wide-angle imaging systems. In *British Machine Vision Conference*.

Gasparini, S., Sturm, P., & Barreto, J. P. (2009). Plane-based calibration of central catadioptric cameras. In *IEEE 12th International Conference on Computer Vision* (pp. 1195–1202).

Geyer, C., & Daniilidis, K. (1999). Catadioptric camera calibration. In *Proceedings of the 7th IEEE International Conference on Computer Vision* (Vol. 1, pp. 398–404).

Geyer, C., & Daniilidis, K. (2000). A unifying theory for central panoramic systems and practical applications. *Proceedings of the 6th European Conference on Computer Vision-Part II* (pp. 445–461). London: UK.

Geyer, C., & Daniilidis, K. (2001a). Catadioptric projective geometry. *International Journal of Computer Vision, 45*, 223–243.

Geyer, C., & Daniilidis, K. (2001b). Structure and motion from uncalibrated catadioptric views. In *Proceedings of the 2001 IEEE Computer Society Conference on Computer Vision and Pattern Recognition, 2001. CVPR 2001* (Vol. 1, pp. I-279–I-286)

Geyer, C., & Daniilidis, K. (2002a). Paracatadioptric camera calibration. *IEEE Transactions on Pattern Analysis and Machine Intelligence, 24*(5), 687–695.

Geyer, C., & Daniilidis, K. (2002b). Properties of the catadioptric fundamental matrix. In *Proceedings of the 7th European Conference on Computer Vision* (pp. 140–154)

Grossberg, M., & Nayar, S. (2001). A general imaging model and a method for finding its parameters. In *Proceedings of the Eighth IEEE International Conference on Computer Vision, ICCV 2001* (Vol. 2, pp. 108–115)

Guerrero, J. J., Murillo, A. C., & Sagüés, C. (2008). Localization and matching using the planar trifocal tensor with bearing-only data. *IEEE Transactions on Robotics, 24*(2), 494–501.

Hansen, P., Corke, P., Boles, W., & Daniilidis, K. (2007). Scale-invariant features on the sphere. In *IEEE 11th International Conference on Computer Vision* (pp. 1–8).

Hansen, P., Corke, P., & Boles, W. (2010). Wide-angle visual feature matching for outdoor localization. *The International Journal of Robotics Research, 29*, 267–297.

Hartley, R. I., & Zisserman, A. (2000). *Multiple View Geometry in Computer Vision.* Cambridge University Press, Cambridge, ISBN: 0521623049.

Heikkila, J., & Silven, O. (1997). A four-step camera calibration procedure with implicit image correction. In *IEEE Computer Society Conference on Computer Vision and Pattern Recognition, 1997* (pp. 1106–1112).

Horn, R., & Johnson, C. (1985). *Matrix analysis.* Cambridge University Press:New York

Horn, R., & Johnson, C. (1991). *Topics in matrix analysis* (2nd ed.). Cambridge: Cambridge University Press.

Jankovic, N., & Naish, M. (2007). A centralized omnidirectional multi-camera system with peripherally-guided active vision and depth perception. In *IEEE International Conference on Networking, Sensing and Control* (pp. 662–667). April 15–17 2007

Kadir, T., & Brady, M. (2001). Saliency, scale and image description. *International Journal of Computer Vision, 45*, 83–105.

Kang, S. B. (2000). Catadioptric self-calibration. In *Proceedings of the IEEE Conference on Computer Vision and Pattern Recognition* (Vol. 1, pp. 201–207).

Kannala, J., & Brandt, S. (2004). A generic camera calibration method for fish-eye lenses. In *Proceedings of the 17th International Conference on Pattern Recognition. ICPR 2004* (Vol. 1, pp. 10–13).

Kosecka, J., & Zhang, W. (2002). Video compass. In *ECCV '02: Proceedings of the 7th European Conference on Computer Vision-Part IV* (pp. 476–490)

Lourenço, M., Barreto, J., & Malti, A. (2010). Feature detection and matching in images with radial distortion. In *IEEE International Conference on Robotics and Automation* (pp. 1028–1034)

Lowe, D. (2004). Distinctive image features from scale-invariant keypoints. *International Journal of Computer Vision, 20*, 91–110.

Matas, J., Chum, O., Urban, M., & Pajdla, T. (2002). Robust wide baseline stereo from maximally stable extremal regions. In *British Machine Vision Conference.*

Mauthner, T., Fraundorfer, F., & Bischof, H. (2006). Region matching for omnidirectional images using virtual camera planes. In *Proceedings of the 11th Computer Vision Winter Workshop 2006, Telc, Czech Republic*(pp. 93–98).

Mei, C., & Malis, E. (2006). Fast central catadioptric line extraction, estimation, tracking and structure from motion. In *IEEE/RSJ International Conference on Intelligent Robots and Systems* (pp. 4774–4779).

Mei, C., & Rives, P. (2007). Single view point omnidirectional camera calibration from planar grids. In *IEEE International Conference on Robotics and Automation* (pp. 3945–3950).

Menem, M., & Pajdla, T. (2004). Constraints on perspective images and circular panoramas. In H. Andreas, S. Barman, & T. Ellis (Eds.), *BMVC 2004: Proceedings of the 15th British Machine Vision Conference.* , London, UK: British Machine Vision Association : BMVA.

Mezouar, Y., Abdelkader, H., Martinet, P., & Chaumette, F. (2004). Central catadioptric visual servoing from 3d straight lines. In *Proceedings. 2004 IEEE/RSJ International Conference on Intelligent Robots and Systems (IROS 2004)* (Vol. 1, pp. 343–348).

Micusik, B., & Pajdla, T. (2003). Estimation of omnidirectional camera model from epipolar geometry. In *Computer Vision and Pattern Recognition* (vol 1, pp. I-485–I-490).

Micusik, B., & Pajdla, T. (2006). Structure from motion with wide circular field of view cameras. *IEEE Transactions on Pattern Analysis and Machine Intelligence, 28*(7), 1135–1149.

Mikolajczyk, K., & Schmid, C. (2004). Scale and affine invariant interest point detectors. *International Journal of Computer Vision, 60*, 63–86.

Miyauchi, R., Shiroma, N., & Matsuno, F. (2007). Development of omni-directional image stabilization system using camera posture information. In *IEEE International Conference on Robotics and Biomimetics (ROBIO 2007)* (pp. 920–925).

Morel, O., & Fofi, D. (2007). Calibration of catadioptric sensors by polarization imaging. In *IEEE International Conference on Robotics and Automation* (pp. 3939–3944)

Murillo, A. C., Guerrero, J. J., & Sagüés, C. (2007). Surf features for efficient robot localization with omnidirectional images. In: *IEEE International Conference on Robotics and Automation (ICRA)* (pp 3901–3907). Rome-Italy.

Orghidan, R., Salvi, J., & Mouaddib, E. M. (2003). Calibration of a structured light-based stereo catadioptric sensor. In *Conference on Computer Vision and Pattern Recognition Workshop, 2003. CVPRW '03* (Vol. 7, pp. 70–70).

Puig, L., & Guerrero, J. J. (2009). Self-location from monocular uncalibrated vision using reference omniviews. In *IROS 2009: The 2009 IEEE/RSJ International Conference on Intelligent Robots and Systems*. St. Louis, MO, USA.

Puig, L., & Guerrero, J. J. (2011). Scale space for central catadioptric systems: Towards a generic camera feature extractor. In *IEEE International Conference on Computer Vision (ICCV)* (pp. 1599–1606).

Puig, L., Guerrero, J. J., & Sturm, P. (2008). Matching of omindirectional and perspective images using the hybrid fundamental matrix. In *Proceedings of the Workshop on Omnidirectional Vision, Camera Networks and Non-Classical Cameras*. Marseille, France

Puig, L., Bermudez, J., & Guerrero, J. J. (2010). Self-orientation of a hand-held catadioptric system in man-made environments. In *IEEE International Conference on Robotics and Automation (ICRA)*, pp. 2549–2555

Puig, L., Bastanlar, Y., Sturm, P., Guerrero, J. J., & Barreto, J. (2011). Calibration of central catadioptric cameras using a DLT-like approach. *International Journal of Computer Vision, 93*, 101–114.

Puig, L., Bermúdez, J., Sturm, P., & Guerrero, J. J. (2012a). Calibration of omnidirectional cameras in practice: A comparison of methods. *Computer Vision and Image Understanding, 116*(1), 120–137.

Puig, L., Sturm, P., & Guerrero, J. J. (2012b). Hybrid homographies and fundamental matrices mixing uncalibrated omnidirectional and conventional cameras. *Machine Vision and Applications, 85*, 1–18.

Ramalingam, S., Sturm, P., & Lodha, S. (2005). Towards complete generic camera calibration. In *Proceedings of the Conference on Computer Vision and Pattern Recognition. CVPR 2005* (Vol. 1, pp. 1093–1098).

Ramalingam, S., Sturm, P., & Lodha, S. K. (2010). Generic self-calibration of central cameras. *Computer Vision and Image Understanding, 114*(2), 210–219. Special issue on Omnidirectional Vision, Camera Networks and Non-conventional Cameras

Scaramuzza, D., Martinelli, A., & Siegwart, R. (2006). A flexible technique for accurate omnidirectional camera calibration and structure from motion. In *Proceedings of IEEE International Conference on Computer Vision Systems*.

Strelow, D., Mishler, J., Koes, D., & Singh, S. (2001). Precise omnidirectional camera calibration. *Proceedings of the Conference on Computer Vision and Pattern Recognition, CVPR 2001* (Vol. 1, pp. I-689–I-694).

Sturm, P. (2002). Mixing catadioptric and perspective cameras. In *Workshop on Omnidirectional Vision* (pp. 37–44). Copenhagen, Denmark.

Sturm, P., & Barreto, J. (2008). General imaging geometry for central catadioptric cameras. In *Proceedings of the 10th European Conference on Computer Vision* (vol. 4, pp. 609–622). Springer.

Sturm, P., & Gargallo, P. (2007). Conic fitting using the geometric distance. In *Proceedings of the Asian Conference on Computer Vision*. Tokyo, Japan: Springer.

Sturm, P., & Ramalingam, S. (2004). A generic concept for camera calibration. In T. Pajdla, & J. Matas (Eds.), *Computer Vision—ECCV 2004, Lecture Notes in Computer Science* (Vol. 3022, pp. 1–13). Berlin, Heidelberg: Springer.

Sturm, P., Ramalingam, S., Tardif, J.-P., Gasparini, S., & Barreto, J. (2011). Camera models and fundamental concepts used in geometric computer vision. *Foundations and Trends in Computer Graphics and Vision, 6*(1–2), 1–183.

Svoboda, T., & Pajdla, T. (2002). Epipolar geometry for central catadioptric cameras. *International Journal of Computer Vision, 49*(1), 23–37.

Swaminathan, R., & Nayar, S. (1999). Non-metric calibration of wide-angle lenses and polycameras. In *IEEE Computer Society Conference on Computer Vision and Pattern Recognition* (Vol. 22, pp. 1172–1178).

Tardif, J.-P., Sturm, P., & Roy, S. (2006). Self-calibration of a general radially symmetric distortion model. In *Proceedings of the 9th European Conference on Computer Vision, Lecture Notes in Computer Science* (vol. 4, pp. 186–199). Springer.

Toepfer, C., & Ehlgen, T. (2007). A unifying omnidirectional camera model and its applications. In *Proceedings of the 11th International Conference on Computer Vision, ICCV 2007* (pp. 1–5).

Tsai, R. (1987). A versatile camera calibration technique for high-accuracy 3d machine vision metrology using off-the-shelf tv cameras and lenses. *IEEE Journal of Robotics and Automation, 3*(4), 323–344.

Vandeportaele, B., Cattoen, M., Marthon, P., & Gurdjos, P. (2006). A new linear calibration method for paracatadioptric cameras. In: *Proceedings of the 18th International Conference on Pattern Recognition. ICPR 2006* (Vol. 4, pp. 647–651).

Vasseur, P., & Mouaddib, E. M. (2004). Central catadioptric line detection. In *British Machine Vision Conference*.

Vedaldi, A. (2007). An open implementation of the SIFT detector and descriptor. Technical Report 070012, UCLA CSD.

Wu, F., Duan, F., Hu, Z., & Wu, Y. (2008). A new linear algorithm for calibrating central catadioptric cameras. *Pattern Recognition, 41*(10), 3166–3172.

Wu, Y., & Hu, Z. (2005). Geometric invariants and applications under catadioptric camera model. *IEEE International Conference on Computer Vision, 2*, 1547–1554.

Wu, Y., Li, Y., & Hu, Z. (2006). Easy calibration for para-catadioptric-like camera. In *IEEE/RSJ International Conference on Intelligent Robots and Systems* (pp. 5719–5724).

Ying, X., & Hu, Z. (2004a). Catadioptric camera calibration using geometric invariants. *IEEE Transactions on Pattern Analysis and Machine Intelligence, 26*(10), 1260–1271.

Ying, X., & Hu, Z. (2004b). Catadioptric line features detection using hough transform. In *Proceedings of the 17th International Conference on Pattern Recognition (ICPR 2004)* (Vol. 4, pp. 839–842).

Ying, X., & Zha, H. (2005). Simultaneously calibrating catadioptric camera and detecting line features using hough transform. *IEEE/RSJ International Conference on Intelligent Robots and Systems, 2005, IROS* (pp. 412–417).

Index

L. Puig and J. J Guerrero, *Omnidirectional Vision Systems*, SpringerBriefs
in Computer Science, DOI: 10.1007/978-1-4471-4947-7, © Luis Puig 2013